AESTHETICS AFTER DARWIN

THE MULTIPLE ORIGINS AND FUNCTIONS OF ARTS

Evolution, Cognition, and the Arts

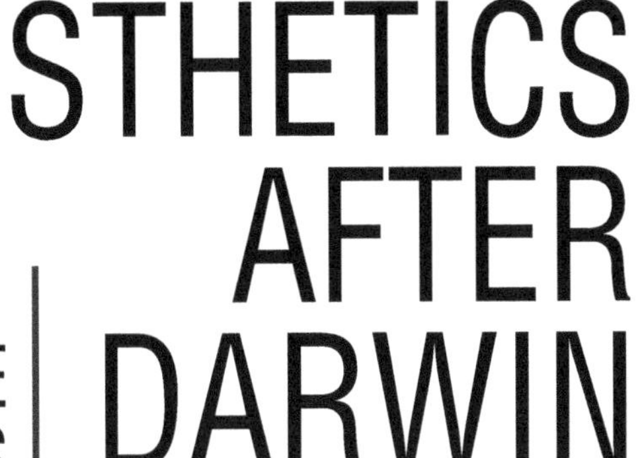

AESTHETICS AFTER DARWIN

THE MULTIPLE ORIGINS AND FUNCTIONS OF ARTS

WINFRIED MENNINGHAUS

Translated by Alexandra Berlina

BOSTON
2019

This research was generously supported by an *opus magnum* grant of the Volkswagen and Thyssen Foundations.

Library of Congress Cataloging-in-Publication Data

Names: Menninghaus, Winfried, author.

Title: Aesthetics after Darwin: The Multiple Origins and Functions of the Arts / Winfried Menninghaus.

Other titles: Wozu Kunst? English

Description: Brighton, MA : Academic Studies Press, 2019. | "The original German version of this book was published in 2011. For the purposes of the present English translation, it was substantially revised." | Includes bibliographical references and index.

Identifiers: LCCN 2019001643 (print) | LCCN 2019012368 (ebook) | ISBN 9781644690017 (ebook) | ISBN 9781644696101

Subjects: LCSH: Arts—Philosophy. | Aesthetics. | Darwin, Charles, 1809-1882. Descent of man.

Classification: LCC BH39 (ebook) | LCC BH39 .M452713 2019 (print) | DDC 700.1—dc23

LC record available at https://lccn.loc.gov/2019001643

ISBN (ebook) 978-1-64469-001-7
ISBN (hardcover) 978-1-64469-000-0

Book and cover design by Lapiz Digital Services.

Published by Academic Studies Press.
1577 Beacon street,
Brookline, MA 02446, USA
press@academicstudiespress.com
www.academicstudiespress.com

Contents

Introduction

Humans have applied ornamental natural objects and artifacts to their bodies and most likely also practiced body painting for some hundred thousand years, if not longer. Extant figurative paintings as well as musical instruments are about forty thousand years old.[1] Given the evolution of the human vocal tract, the presumably oldest human arts, those of singing (and perhaps also of concomitant dancing), may well have been practiced already for several hundred thousand years. However, by their very nature, these pure performance arts left no material traces, and hence no conclusive empirical proof, in the archeological record. In any event, countless animal practices of artful singing, dancing, and multimedia displays are likely to be far older than their human analogues. Importantly, these animal practices, like those of humans, typically involve extensive learning and practicing, and accordingly show substantial "cultural" variation.[2] In fact, Darwin exclusively spoke of animal "arts" in cases where the respective practices are *not* fully genetically coded but require ontogenetic learning.[3]

Findings and hypotheses of this sort constitute the rationale for the questions asked by evolutionary aesthetics: Can the animal and the human arts both be conceived as evolved adaptations? And has their respective evolution potentially been driven by similar functional benefits? Moreover, looking at features distinctive of the human arts only (such as the use of

---

1 For details, see below, p. 92–101.

2 Kroodsma, "Learning and the Ontogeny of Sound Signals in Birds"; Payne et al., "Biological and Cultural Success of Song Memes in Indigo Buntings." For more details, see below, p. 4–5, 32–34.

3 Darwin, *The Descent of Man, and Selection in Relation to Sex*, I 55–56. Quotations from this book follow the Princeton 1981 edition, which reproduces the text of the first edition of 1871. The Roman numeral I or II denotes the relevant part of the book, the Arabic number the page. Quotations from the second edition (1874) of Darwin's book, which contains some amendments and additions, are pointed out by footnotes.

technology for art production purposes, fictional content, etc.) raises the question of how previously evolved human adaptations facilitated their emergence? And do the current human arts still show traces of their potential ancient origin and functions?

The present book discusses theoretical hypotheses regarding these fundamental questions, while reviewing the available empirical evidence. Evolutionary explanations of the animal and human arts in the end all involve a response to the question "Wherefore art?"

After all, according to the standard theory, evolutionary processes favor some bodily and behavioral traits over others, if and to the extent they provide some advantage in a given—or coevolving—evolutionary environment. Functional hypotheses regarding the evolution of the human arts are tricky battlegrounds.[4] As already indicated, archaic functions of bodily and behavioral adaptations—and hence the very drivers of their evolution according to the standard model—have often left no archeological traces. If they did, these traces are typically fairly difficult to read. In any event, it is far from clear to what extent, if at all, the function of a behavioral trait observed today is identical to the archaic function that originally drove its evolution.[5] Regarding archaic practices of the human arts, the worst case applies: visual artworks and musical instruments (but not music itself) are handed down to posterity, but without any cues as to their meaning, uses, and functions in cultures from which nothing is left but material traces limited to particularly durable objects. Comparisons with historically documented and current human cultures offer some guidance for arriving at hypotheses regarding the role of the arts as practiced tens if not hundreds of thousands of years before our time; yet again, it can by no means be taken for granted that the functions of the human arts have not significantly changed over such a long stretch of time.

For all these fully acknowledged risks and difficulties, the first three chapters of the present book, while always including empirical research into physiological mechanisms, discuss primarily three hypotheses of function. The first two are frequently discussed as potential evolutionary functions:

4 Cf. Tinbergen, "On Aims and Methods of Ethology," and Fitch, "The Biology and Evolution of Music," 174 sqq.

5 Cf. ibid., 175.

1. The human arts of singing and self-adornment evolved as competitive practices of aesthetically appealing self-presentation (self-advertisement) in the context of *sexual courtship*. This is how Charles Darwin's treatment of the animal and the human arts is typically understood. Chapter 1 discusses this hypothesis with an emphasis on the limitations of the human–animal parallel that Darwin explicitly highlighted. Many scholars have neglected to discuss the limitations of this parallel. The role of the arts for *social self-distinction* can be considered as an extension of this hypothesis beyond the narrower sexual context.
2. The arts of singing, dancing, and ritual multimedia displays evolved as communal practices of reinforcing *social cooperation and cohesion* within or between human groups. This hypothesis has some advocates in evolutionary theory and enjoys much support in ethnology and social anthropology. It is discussed in Chapter 2.
3. The arts serve the *ontogenetic self-(trans)formation of individuals*. This hypothesis by and large underlies the humanist concept of "liberal education." Because this hypothesis is not a hypothesis about the evolutionary origins of art, but about the role of art education in modern school systems and societies, Chapter 3 does not treat it at length, but exclusively focuses on the question: How incompatible or compatible is this hypothesis with the hypotheses of sexual competition/social distinction and social cooperation/cohesion?

The first and longest chapter of the present book follows in the footsteps of the giant of evolutionary theory, Charles Darwin. Surprisingly, many important theoretical distinctions Darwin proposed for an evolutionary understanding of the visual, vocal, and verbal human arts have gone completely untreated not only in textbook renditions of his theory but also in the narrower context of evolutionary theory. Put briefly, the reason for this lacuna lies in the strongly asymmetrical reception of Darwin's theory of *natural selection*—as presented in his book *The Origin of Species* (1859)—and his theory of special processes of *sexual selection* (*The Descent of Man, and Selection in Relation to* Sex, 1871). Whereas the theory of natural selection covers all interactions between an organism and its ecological niche at large (including climate, food resources, competition with other species, self-imposed changes on the ecological niche), Darwin's theory of sexual selection primarily if not exclusively addresses within-species competition for and choice of sexual partners.

Darwin hesitated many years before he finally decided to set the processes of sexual selection categorically apart from those of natural selection. In the end, he did so for two reasons. First, sexual selection largely suspends the otherwise pervasive rule of violence and fight. Males do not automatically win over females by outfighting other males; rather, they must in addition court the females in a nonviolent way by the charms of their appearance, songs, dances, or multimedia performances. Moreover, many of the bodily ornaments that are selectively useful for sexual courtship are fairly disadvantageous in other contexts of natural selection; for instance, they often reduce capacities for flight or make the individual highly visible for predators. Hence winning over by beauty seems to involve special processes that often appear to be at odds with the mechanisms driving natural selection and are to this extent "autonomous" from the more pragmatic concerns of daily life.

Darwin's contemporaries strongly rejected his proposal of special processes of sexual selection—not least because it attributes to the female sex the "power" not only of sexual choice but also of driving the evolution both of the male body and of male behavior. Neo-Darwinist theories of the twentieth century likewise mostly rejected the frontal assault that Darwin's theory of the evolution of sexual ornaments, beauty competitions, and sexual singing entailed for the role of the male sex.[6]

There is a second reason why Darwin's book *The Descent of Man, and Selection in Relation to Sex* is yet to be discovered by both humanists and scientists. Humanists do not read this book, because they are not interested in biological evolution, and scientists do not read it, because they do not read books of more than 800 pages anymore, all the less so if they date back some 150 years. In fact, several evolutionary biologists I spoke with readily confessed to never having read Darwin's book. As a result, the prevailing notion regarding its content is based on abstracts that made it into textbooks, and these do not entail what will be brought to the fore in the upcoming pages. Even a recent book wholly based on Darwin's theory of sexual selection skips over many of Darwin's finer distinctions the present study places a special focus on. As though he needed some excuse that he actually read Darwin's book, the author starts out with a remark informing the reader that s/he should not be concerned about getting too close a

6 For a more detailed account of the unhappy fate of Darwin's theory of sexual selection and the arts, see Prum, *The Evolution of Beauty*, 17–53.

reading of Darwin's own words: "It's not that I am interested in doing a Talmudic-style investigation of Darwin's every word."[7]

If some readers feel that the first chapter of the present book does precisely this, that's fine with me. An integral part of this effort of close reading is to bring out something that to date has been completely disregarded by both the sciences and the humanities: Darwin consistently builds bridges between his evolutionary hypotheses regarding the "sense of beauty" and humanist traditions of aesthetics. Specifically, his general assumptions concerning aesthetic virtues and aesthetic judgment are strongly informed by British and German eighteenth- and nineteenth-century philosophical aesthetics and ethnology. Even more importantly, close attention to the many nuances of Darwin's considerations reveals that he uses the hypothesis of function underlying his animal model of the arts—displays of beautiful looks and artistic capabilities enhance success in competitive sexual courtship—only with great caution and decisive qualifications in his quest for an evolutionary understanding of the human arts. In the end, this understanding will be shown to be anything but the simple "singing for sex" hypothesis to which it has been largely boiled down.

A Cooptation Account of the Evolution of the Human Arts

Chapter 4 of the present book pursues a different avenue toward an evolutionary theory of the human arts. Here, the primary focus is not on the functional attractor(s) that hypothetically drove the evolution of the human arts. Rather, the focus is on *how* the human arts may have evolved and specifically on how previously evolved adaptations may have paved the way for the evolution of the human arts. Both Darwin and recent evolutionary theory entail provisions for evolutionary processes that need not proceed via the selection of a special and task-specific novel adaptation. Rather, a new behavior can just as well evolve via the "cooptation"[8] of existing traits—be they task-specific adaptations or nonadaptive side effects of adaptations—for novel (additional) functions.

As Gould and Vrba point out,[9] already Darwin hinted at several examples for such evolutionary cooptations, specifically regarding complex traits of human behavior.[10] This type of evolution relies on cross-modular,

7 Ibid., 17.
8 Gould and Vrba, "Exaptation," 13.
9 Ibid., 5 sqq.
10 See, for instance, Darwin, *Descent of Man*, II 335.

or domain-general, uses of our cognitive, emotive, and behavioral abilities and dispositions and hence accords with the widely shared notion that the peculiar creativity and flexibility of the human mind specifically manifests itself in such cross-modular uses of its special evolved capabilities.[11] Ontogenetic learning and purely cultural transmission play a great role in such evolutionary processes. Some models of evolution even completely rely on nongenetic processes of feedback and transmission.[12]

Specifically, the cooptation model of the human arts outlined in Chapter 5 proposes that humans extended the field of the arts far beyond the hypothetical analogues of the bird arts (i.e., the singing and dancing that Darwin assumes in the ancestors of the modern Homo sapiens) by coopting distinctly human capacities for play behavior, tool use, and symbolic cognition, which developed independently from the arts, into the field of art production and reception. For birds, singing or dancing for sex is a very serious predicament; it is not inherently playful or exploratory. The human arts, by contrast, for all the passion and seriousness with which they, too, are pursued, often do entail elements of play and exploration that are relatively free from direct pragmatic consequences. Together, play behavior, technology, and the cooptation of symbolic cognition not only extended the range of the hypothetical arts of aesthetically appealing self-presentation and (sexual and social) display; these distinctly human capacities also added important new cultural levels and functions.

The original German version of this book was published in 2011. For the purposes of the present English translation, it was substantially revised.

11 Cf. the reflections on our higher cognitive capacities in Fodor, *The Modularity of Mind*, Donald, "Art and Cognitive Evolution," 17, and Mithen, *The Prehistory of the Mind*. See also Neumann, *Funktionshistorische Anthropologie der ästhetischen Produktivität*, particularly 117.

12 See, for instance, Heyes, "Grist and Mills: On the Cultural Origins of Cultural Learning," and the entire volume on the evolution of human cognition of which this article is part.

1

Competitive Courtship and Aesthetic Judgment/Choice: Darwin's Model of the Arts

Darwin's evolutionary aesthetics primarily explains the evolution of aesthetic preferences for particular features of bodily looks and their role in competitive sexual courtship and choice. His theory of the evolution of the arts of singing, dancing, and poetic speech is an intriguing extension of this model of "sexual selection" for beautiful looks, one that has surprising observations and interpolations in store to this very day.

Section 1 analyzes Darwin's general assumptions regarding features of aesthetic appeal, mechanisms of beauty judgments, and their role in sexual courtship and choice. Section 2 is devoted to Darwin's visual aesthetics. It traces the reasons why Darwin regarded naked skin as the single most important and most distinctive human body "ornament" (2.1), discusses how this very special ornament favored the development of the human arts of self-painting and other types of visual self-decoration (2.2), and reveals the importance of naked skin and of the correlative practices of wearing and putting on clothing for the emergence of visual "imagination" (2.3). Section 3 discusses Darwin's theory of the musical arts (3.1) as well as of music-elicited emotions (3.2), and comments on his hypotheses regarding the poetic and rhetorical elaboration of language in light of both classical rhetoric and recent research in rhetoric and poetics (3.3). Section 4 discusses the merits and limits of the new perspectives on the human arts that Darwin brought to the fore by comparing them with the performances of peacocks and songbirds.

1. The "Sense of Beauty": Darwin's General Assumptions regarding Aesthetic Virtues and Aesthetic Judgment

Darwin spent decades ruminating about the "beauty" of bodily looks, as it represented a major challenge to his theory of natural selection. Decorative feathers, horns, and antlers have reached astonishing forms and sizes in many animals, so much so that they are a massive handicap "in the general conditions of life"[1] and mostly of little use as weapons. A cardinal text of philosophical aesthetics, one that Darwin cites more than once, seems to have directly guided his theorizing regarding a "sense of beauty"[2] or "taste": Edmund Burke's *A Philosophical Enquiry into the Origin of Our Ideas of the Sublime and Beautiful* (1756). Burke already conceptualized the extreme beauty of the peacock's ornaments in terms of a conflict with practical "fitness," and specifically the "aptitude" for flying.[3] How could the peacock's tail and analogous "ornaments" evolve despite the fact that they impair movement and flight capacities, renders the respective animals highly visible to predators, and hence seem irreconcilable with Darwin's model of adaptive natural selection?

Responding to this challenge, Darwin ended up proposing a rich set of general assumptions regarding aesthetic virtues, aesthetic judgment, and their role in sexual courtship and choice. The key concepts he relies on have a long tradition before him and keep informing research in aesthetics to date: beauty, taste, novelty, familiarity, prototypicality, exaggeration, variety for the sake of variety, caprice.

To start with, Darwin holds that the first step toward the evolution of a preferred body ornament is generally arbitrary. Some existing feature grows slightly more expressed through (genetic) variation. The very novelty, relative rarity, and differential quality of this "exaggeration" can attract attention. This assumption is in line both with classical aesthetics and with the current understanding of the brain functions of "predictive coding,"[4] of consistently comparing all incoming information with anticipations based on prior knowledge. If anticipations are met, no special response is required; if minor detours from expectations occur, the brain automatically devotes more attention and more effort to dealing with the not fully expected stimulus.

---

1 Darwin, *Descent of Man*, I 279.
2 Darwin, *Descent of Man*, I 63–65.
3 Burke, *A Philosophical Enquiry*, 106.
4 Cf. Friston, "The Free-Energy Principle: A Rough Guide to the Brain?"

Which particular characteristics of bodily appearance end up becoming preferred as ornamental traits and pushed to ever higher degrees cannot be predicted by Darwin's theory; random variation plays a great role in this context. At the same time, some mechanisms act as constraints on potential aesthetic preferences for random variations. Thus, preferred ornaments should *not* be similar to those of closely related species,[5] and they should, of course, be in line with the general sensory dispositions of a species. In any event, since sexual ornaments mostly lack a pragmatic function outside the context of sexual courtship, all they need to do is to *impress and please as such*—and this can, in principle, be achieved in many ways. This is why Darwin's model postulates a great variance and arbitrariness of features preferred for their "beauty." The "caprices of fashion"[6] are a cultural analogue Darwin refers to on more than one occasion.

The over-expression of ornamental traits need not be large in order to attract attention and, potentially, desire. Small, even minimal differences suffice. Thus, there is no conflict between the preferences for "mere novelty," "change for the sake of change" and "mere variety"[7] and for the equally well-established preference for the average prototype of a species and hence for familiarity.[8] Pointing to male preferences regarding women's looks, Darwin exemplifies this combination by stating on the one hand: "The men of each race prefer what they are accustomed to behold," and on the other: "As the great anatomist Bichat long ago said, if everyone were cast in the same mould, there would be no such thing as beauty. If all our women were to become as beautiful as the Venus de Medici, we should for a time be charmed; but we should soon wish for variety; and as soon as we had obtained variety, we should wish to see certain characters in our women a little exaggerated beyond the then existing common standard."[9]

This dual preference for norm conformity[10] and some degree of deviation (novelty, variety, exaggeration) corresponds exactly to a basic Aristotelian rule for the verbal arts. According to this rule, the poet should include something foreign (*xenón*) and exaggerated (*aúxesis*) in his diction,

5 Eberhard, *Sexual Selection and Animal Genitalia*; Lande, "Genetic Correlations between the Sexes in the Evolution of Sexual Dimorphism and Mating Preferences."

6 Darwin, *Descent of Man*, II 230, 339.

7 Darwin, *Descent of Man*, II 230.

8 Cf. Kant, *Critique of the Power of Judgment*, § 17; Fechner, *Vorschule der Ästhetik*, 262; Martindale and Moore, "Priming, Prototypicality, and Preference"; Winkielman et al., "Prototypes Are Attractive."

9 Darwin, *Descent of Man*, II 354.

10 Cf. Langlois et al., "What Is Average and What Is Not Average about Attractive Faces," and Perrett, "Facial Shape and Judgments of Female Attractiveness," 239–42.

yet also remain close to the usual and the familiar. The deviating elements provide a pleasure that is based on astonishment (*hedy dè tó thausmastón*); the familiar, on the other hand, is pleasant because it is easy to process. The two are to be interwoven in an appropriate fashion, according to context and genre.[11] A recent theory of "optimal innovation" in the field of aesthetic and rhetorical forms states the same,[12] and so does a theory of aesthetic pleasure based on the "predictive coding"-hypothesis regarding the working of the human brain.[13] Continued preference for minimal deviations can lead to a self-reinforcing and, evolutionarily speaking, relatively rapid process of increasing nuances of difference step by step;[14] in the end, this dynamic can result in extreme expressions of body ornaments that guide sexual choice. The notion of a "runaway selection" has been largely accepted since Ronald A. Fisher's mathematical reformulation of Darwin's model.[15] (Extremely hypertrophying male and female features of looks through self-shaping or plastic surgery are intentional and non-evolutionary analogues to this process.)

In the auditory and visual arts, Darwin saw the same mechanisms at work as in the preference for certain body ornaments. Self-evidently, the arts can satisfy the demand for novelty, variety, and exaggeration much faster than the evolution of natural differences between sexual ornaments of the body. This applies already to the animal arts of singing and dancing: the more they depend on ontogenetic practicing and learning, the better they serve the demand for differences in performance. Darwin therefore used the concept of "art" for animal performances exclusively in cases when they involve substantial individual efforts of refining the singing and dancing,[16] and hence involve a *dual inheritance* of a phylogenetic and an ontogenetic nature.[17] Ever since Darwin, substantial evidence has been collected regarding the importance of learning and local traditions, and hence of nongenetic change, in the animal arts, specifically in birds.[18] Ontogenetic

11 Aristotle, *Rhetoric* 1404a.

12 Giora, "Weapons of Mass Distraction: Optimal Innovation and Pleasure Ratings," 115–41. Cf. also Miller, "Evolution of Human Music through Sexual Selection," 344–46.

13 Van de Cruys and Wagemans, "Putting Reward in Art: A Tentative Prediction Error Account of Visual Art."

14 Darwin, *Descent of Man*, II 351.

15 Fisher, *The Genetical Theory of Natural Selection*.

16 Darwin, *Descent of Man*, I 55–56.

17 Cf. Verpooten, *Art and Signaling in a Cultural Species*; Boyd, *On the Origin of Stories*; Boyd and Richerson, *Culture and the Evolutionary Process*.

18 Cf. Kroodsma, "Learning and the Ontogeny of Sound Signals in Birds"; Kroodsma, *The Singing Life of Birds*; Payne et al., "Biological and Cultural Success of Song Memes in

factors, general learning abilities, and local traditions thus gain importance compared to a purely genetic variation between sexual body ornaments, which is the only object of Fisher's reformulation of Darwin's model.

Darwin attributed the discovery of the "principle" of self-reinforcing preferences for particular ornaments to Alexander von Humboldt.[19] Humboldt's travel reports provided many of the ethnic examples for both the natural and cultural body fashions that Darwin refers to. These examples concern primarily the interaction between natural body features and their cultural treatment, such as the amplification of biologically evolved differences in skin color through cultural body painting.[20] For Darwin, the differential treatment of the male beard is a particularly conspicuous example for such a feedback loop between nature and culture:

> It is remarkable that throughout the world the races which are almost completely destitute of a beard dislike hairs on the face and body, and take pains to eradicate them. The Kalmucks are beardless, and they are well known, like the Americans, to pluck out all straggling hairs; and so it is with the Polynesians, some of the Malays, and the Siamese. . . . On the other hand, bearded races admire and greatly value their beards; among the Anglo-Saxons every part of the body, according to their laws, had a recognized value; "the loss of the beard being estimated at twenty shillings, while the breaking of a thigh was fixed at only twelve."[21]

Analogously, if the biological evolution has resulted in more oval or in more broad faces, each respective preference is then culturally amplified.[22] Capricious practices that make small feet even smaller than they are naturally[23] or stress broad female behinds,[24] elaborate hairdos,[25] or different fashions for teeth and head forms[26] also fit well into Darwin's series of examples along the lines of the Humboldt principle.

Even more strikingly, many animal species probably evolved in the first place because deviating preferences for sexually attractive ornaments finally

---

Indigo Buntings"; Saranathan et al., "Genetic Evidence Support Song Learning"; Prum, "Coevolutionary Aesthetics in Human and Biotic Artworlds," 826–27.

19 Darwin, *Descent of Man*, II 351.

20 Darwin, *Descent of Man*, II 346–47, 352.

21 Darwin, *Descent of Man*, II 349.

22 Darwin, *Descent of Man*, II 344–45, 354.

23 Darwin, *Descent of Man*, II 352.

24 Darwin, *Descent of Man*, II 345–46.

25 Darwin, *Descent of Man*, II 340, 348.

26 Darwin, *Descent of Man*, II 340–41.

led to the emergence of separate subspecies.[27] In this regard, Darwin points to birds that are otherwise of the same species yet have developed opposite extremes of coloration, such as black and white swans, storks, or ibises.[28]

The tendency to intensify sexual ornaments can thus lead to pronounced aesthetic *contrasts* between species. In general, such self-reinforcing preferences, once they take hold, can evolutionarily favor very striking forms, colors, and movement/display patterns in the long run—unless natural selection stops this trend and/or as long as it is supported by the respective animal's sensory dispositions in combination with the light and visibility conditions of its ecological niche. Hence the many colorful, indeed garish and eccentric examples that Darwin presents in his panorama of the natural addiction to ornamentation. The preference for strong contrasts, for pure, luminous colors, and for elaborate aesthetic displays has meanwhile been confirmed in studies on animals and humans (children in particular).[29] At the same time, it is just as clear that human cultural aesthetics regarding paintings, clothes, and interior design can likewise favor the very opposite tendency (i.e., preferences for subtle color contrasts).

Darwin's examples and his general assumption regarding the interaction of aesthetic preferences and sexual choice keep stressing that, at least partially, the aesthetic "sense of beauty" or "taste for the beautiful"[30] has its own laws that cannot be (fully) derived from pragmatic concerns and functions. Such remarks are well in line with the modern notion of "aesthetic autonomy." Moreover, locally divergent aesthetic preferences within a species can hardly serve as genetically determined fitness indicators, as demonstrated by the beards of human males (unless it can be demonstrated that a beard and its absence respectively correlate with "good genes" in the different populations).

Kant and idealist aesthetics combined the hypothesis of beauty as being devoid of concept and purpose with attributing to it important *functions* for human cognitive and affective needs.[31] Darwin's functional hypothesis

---

27 Cf. Gould and Gould, *Sexual Selection*, 89 sqq.

28 Darwin, *Descent of Man*, II 230–31.

29 Cf. Fechner, *Vorschule der Ästhetik*, 231–34, and Rensch, "Ästhetische Grundprinzipien bei Mensch und Tier." The latter study, however, finds a preference for pure colors only in some animals. Cf. also Rensch, "Ästhetische Faktoren bei Farb- und Formbevorzugungen von Affen"; Tigges, "Farbbevorzugungen bei Fischen und Vögeln"; Ryan, "Sexual Selection, Sensory Systems, and Sensory Exploitation"; Miller, "Evolution of Human Music through Sexual Selection," 341–43.

30 Darwin, *Descent of Man*, I 63–64.

31 Cf. Kant, *Critique of the Power of Judgment*, § 1, 10; Schmücker, "Funktionen der Kunst," particularly 15–20; and Menninghaus, *Kunst als "Beförderung des Lebens."*

regarding the many "capricious" body ornaments of natural species focuses precisely on the very sexual implications from which idealist aesthetics tried to purge the concept of "pure" aesthetic pleasure. These ornaments, Darwin holds, provide a competitive advantage in the highly specialized context of sexual courtship. The tendency to disregard adverse effects of beautiful ornaments on other aspects of everyday life makes up the "autonomy" of the ornament; its adaptive benefits in contexts of sexual courtship and choice make up its functional value.

To be sure, Kant's hypothesis of a "disinterested" aesthetic judgment—in which we still may take some "intellectual interest"[32]—is nearly reversed in Darwin's theory. After all, for Darwin, a marked function for success in competitive sexual courtship is ultimately driving the feats of ornamentation. Still, in the situation of being courted, peahens prefer a more splendid male ornament just as immediately and conceptless as required by Kant's model. They do not prefer the most beautiful peacock because they imagine him to be most useful for the proliferation of their own genes. In other words, the *proximate* mechanisms of aesthetic sexual courtship and choice should not be confused with the *ultimate* mechanism underlying this behavior (i.e., its effect on self-reproduction).[33] Rather, *the displays of aesthetic virtues in bodily appearance, song, and dance and the responses to these displays in birds and other animals are special adaptations for the phase of sexual courtship and choice only—and hence for showing/producing and evaluating aesthetic virtues—but not for their possible consequences (copulation, pregnancy, rearing the young).* The important theoretical distinction evolutionary theory makes between proximate and ultimate mechanisms thus implies that the aesthetic appreciation of courtship displays is driven by its own specialized and self-contained mechanisms. It is thus categorically different from the adaptations for sexual copulation, regardless of the potential temporal and even causal link between the two.

On a similar vein, human artists typically strive for some attention and acceptance on the part of a wider or narrower audience. This may in some cases imply that appreciation for their art also enhances their social opportunities and even their sexual success. Still, such consequences would not mean that their artistic efforts are primarily driven by pragmatic goals and interests rather than by inherent artistic motivations. Therefore, the proximate mechanisms of artistic production need to be categorically distinguished from the potential functions they might serve (or not serve) in the end.

32 Kant, *Critique of the Power of Judgment*, § 42.

33 For a similar argument, see Davies, *The Artful Species*, 13–14.

The failure to differentiate between proximate mechanisms of aesthetic appreciation and potential ultimate functions for sexual purposes has led to several confusions and pseudo-alternatives. The most widespread among these is the idea that Darwin's evolutionary aesthetics does not deal with "beauty" at all, but only with physical "attractiveness."[34] To use Kant's terms, "disinterested pleasure" is at the basis of "purely" aesthetic perception and judgment; however, judging the artistic displays of sexual courtship—the argument goes—is not disinterested. This argument does not only fail to question the historic specificity of the modern decree of the disinterestedness of aesthetic pleasure, but also it fails to acknowledge that several languages and philosophical theories from antiquity until today do not hesitate to designate the sexually attractive appearance of men and women as "beauty." Some philosophers, and particularly Kant, tried to banish even the most subtle affinity to sensuous and sexual charms from the lofty realm of transcendental aesthetics. Thus, "highbrow" culture—most notably, eighteenth-century German bourgeois culture—attempted to draw a rigid line of demarcation between the genuinely "aesthetic" domain of beauty and anything even remotely sexually attractive. British and French treatises of aesthetics have not advocated a similarly rigid purging, if not exorcism, of all sexual implications from the very concept of beauty.

Speaking of sexual body "ornaments," Darwin connects evolutionary biology to a central category of philosophical aesthetics. Since the later eighteenth-century aesthetics, ornaments have been the epitome of purposeless and nonconceptual beauty. This is why arabesques, ornamental vines, and other parerga dominate among the examples of "free" beauty Kant provides in his *Critique of the Power of Judgment*.[35] Owen Jones's *Grammar of Ornament* (1856) offered Darwin a contemporary panorama of the aesthetics of ornament. Darwin pictures the evolution of certain body parts in a fashion fully analogous to the application of ornaments. This corresponds to the biological fact that many body ornaments only become prominent in times of sexual courtship and then disappear, immediately or gradually. Such ornaments are no longer conceived as the work of a Creator who has not only created nature wisely but also adorned it beautifully. Instead, sexual selection is considered the driving force: over very long periods of time, countless individual acts of "choice" by the opposite sex made certain "ornaments" hereditary and more pronounced, thereby translating into a genuine power of evolutionary "selection." Darwin is the

34 For instance, Eibl, *Kultur als Zwischenwelt*, 159.

35 Cf. Kant, *Critique of the Power of Judgment*, § 16, and Menninghaus, *In Praise of Nonsense*, 72–92.

first and maybe the only author in the history of aesthetics who stringently conceived of bodily beauty in terms of an aesthetics of ornaments.

Darwin's concept of "caprice"[36] emphasizes the arbitrariness, if not the near absurdity, of elaborate decorative traits when viewed from a pragmatic perspective. To Darwin, the capricious preferences for exuberant plumage are in the final analysis as erratic as the predilection that led to the development of pink backsides in certain monkeys.[37] Considered as quirks, whims, and arbitrary hobbyhorses, these caprices have something endearing, and at the same time something weird, even slightly mad. German equivalents of "caprice" are *Laune*, *Marotte*, *Tick*, and *Manier*.[38] Kant's definition of an aesthetically "capricious manner [*launichte Manier*]"[39] is part of the tradition that informed Darwin's use of the term.

Darwin's notion of a "capriciousness of taste" is also closely reminiscent of a literary phenomenon, namely, the equally capricious poetics of digression, as exhibited in Lawrence Sterne's *Tristram Shandy* (1759–67) and its German Romantic relatives. Sterne's novel promises to offer the "Life and Opinions" of its protagonist; however, due to constant interruptions of the expected storyline through diverse digressions, multiple self-proliferating excursions of a theoretical, historical, and narrative nature, "capricious" anecdotes, and frame stories, it takes *Tristram Shandy* several hundred pages to get as far as the protagonist's birth. The narrative art of German Romanticism also makes ample use of such digressions and techniques of self-proliferating arabesques. Just as in Sterne's novel, multiple paratexts and addenda keep forcing themselves into the foreground, to the point of obstructing the expectable unfolding of the core narrative.

From Friedrich Schlegel via Edgar Allan Poe to Charles Baudelaire, such literary practices were technically designated by reference to a specific type of visual ornament, namely, the "arabesque," with the rococo arabesque being of particular importance.[40] The concept of the arabesque thrived in music, too, above all in the form of "capriccios." Talking about the caprices of taste and the extravagances of ornamental fashion, Darwin is alluding to this rich cultural horizon. This is more than a metaphor: the plumage of many a bird of paradise could quite literally work as a rococo ornament. Thus, Darwin uses the same categories for natural bodies that were used

36 Darwin, *Descent of Man*, II 230, 339.

37 Darwin, "Sexual Selection in Relation to Monkeys," 207.

38 Cf. Strasser, *Bedeutungswandel*, and Menninghaus, *In Praise of Nonsense*, 15–31.

39 Kant, *Critique of the Power of Judgment*, § 54.

40 Cf. Oesterle, "Vorbegriffe zu einer Theorie der Ornamente"; and Menninghaus, *In Praise of Nonsense*, 72–159.

within historical aesthetics to describe phenomena that ostentatiously display their artificiality, such as rococo arabesques and the provocative writings of Lawrence Sterne, Ludwig Tieck, or E. T. A. Hoffmann.

Most other terms Darwin uses to characterize all sorts of aesthetically preferred stimuli (bodily features, songs, dances, and multimedia performances) are likewise derived from philosophical aesthetics and rhetoric (novelty, variety, rarity, exaggeration).[41] However, he uses these categories in a highly innovative fashion. Concepts that have been previously applied to phenomena of art and decoration are adopted for purposes of an unprecedented aesthetics of natural bodies. The connection between aesthetic preference and sexual selection makes "aesthetic judgment," which has a purely *reflective* character in Kant's aesthetics,[42] directly *constitutive* for the very existence, and even for the gradual exaggeration of its objects (i.e., the sexual body ornaments). Projected back onto conventional aesthetics, Darwin bestows the role of aesthetic judgment—which is typically exerted by the female sex that evaluates the looks and/or presentations of male competitors—with an enormous power: female choice[43] does not only *evaluate* given bodily ornaments and artistic presentations, but literally *drives* the future evolution of these traits (i.e., of ever more refined physical ornaments and display practices of great splendor, and hence of "new art").

Experiments suggest that animals, just like humans, tend to prefer exaggerated, supernormal stimuli, and that a preference, once established, can rapidly propel itself to extreme levels (*peak shift effect*).[44] For instance, through food rewards rats were conditioned to prefer squares to other geometric forms. In the next step, a non-square rectangle was introduced and associated with an even larger reward than the square. As expected, the rats learned to reliably prefer the rectangle. Less predictable was the third part of the experiment. The rats were offered the opportunity to choose between the rectangle they already knew and associated with large rewards and another rectangle, the proportions of which were even more different from those of a square. Interestingly, rats picked this novel variant, without undergoing any reward-based conditioning in favor of it. A possible explanation is thus that they chose the larger difference from the original square (i.e., the exaggeration of *non-squareness*).

---

41 Cf. Darwin, *Descent of Man*, I 63–65, II 230h, 339, 351, 354. Bredekamp supposes that Darwin's concept of "variety" was influenced primarily by William Hogarth's *The Analysis of Beauty* (1753); cf. Bredekamp, *Theorie des Bildakts*, 314 sqq.

42 Cf. Kant, *Critique of the Power of Judgment*, § IV.

43 Darwin, *Descent of Man*, II 273.

44 Ramachandran, "The Science of Art," 18.

This experimental finding supports Humboldt's and Darwin's principle of an inherent tendency to push specific select characteristics to ever higher degrees. According to these assumptions, just as in Kant's theory of aesthetic judgment, there can be no stable objective standard of beauty.[45] Quite the opposite: Darwin stresses the greatly varying outcomes the process of "capriciously" preferring initially arbitrary characteristics can lead to. These characteristics can both be exaggerated ever further, but also go out of fashion at some point and can even be replaced by their very opposite:[46]

> Many of the faculties, which have been of inestimable service to man for his progressive advancement, such as the powers of the imagination, wonder, curiosity, an undefined sense of beauty, a tendency to imitation, and the love of excitement or novelty, could hardly fail to lead to capricious changes of customs and fashions.[47]

Here, too, Darwin is inspired by Alexander von Humboldt:

> "If painted nations," as Humboldt observes, "had been examined with the same attention as clothed nations, it would have been perceived that the most fertile imagination and the most mutable caprice have created the fashions of painting, as well as those of garments."[48]

Darwin regards cultural "fashions"[49] as a continuation and substitute of the changes undergone by natural bodies due to the capricious traits favored by sexual selection. Of course, cultural fashions radically increase the speed with which sexually preferred ornaments can change, compared to bio-evolutionary selection. In any event, the great role both of familiarity and of unpredictable, seemingly arbitrary ("capricious") changes makes it crystal clear that for Darwin the highly general, or even biologically evolved, principles of the "sense of beauty" by no means imply that they should lead to the same result in all cultural contexts. After all, preferences for minor "capricious" changes can lead in very different directions, and once the baseline of ontogenetic familiarity and mere exposure differs between cultural contexts, the working of the preferences for ongoing minor changes can also not yield converging results. Hence universal principles

---

45 Cf. Darwin, *Descent of Man*, II 353–54.
46 Darwin, *Descent of Man*, II 338–54.
47 Darwin, *Descent of Man*, I 64.
48 Darwin, *Descent of Man*, II 339.
49 Darwin, *Descent of Man*, II 230–31, 340, 352.

of the "sense of beauty" and great cultural variance are not antithetical for Darwin, and he relishes in reporting in great detail a rich panorama of ethnic and cultural differences in preferred appearance as well as in practices of self-ornamentation and self-shaping.[50] The way he conceives of the "sense of beauty" positively predicts rather than excludes great ethnic, cultural, historical, and even individual differences, thereby defying topical assumptions which consider an evolutionary biology of the "sense of beauty" as irreconcilable with such differences.

The analogy between the capriciousness of sexually preferred traits and cultural fashions was obviously striking for Darwin and his contemporaries. In 1872, George H. Darwin, a relative of Charles Darwin, opened an essay on "Development in Dress" with the following hypothesis: "The development in dress presents a strong analogy to that of organisms, as explained by the modern theories of evolution."[51] Herbert Spencer extended this evolutionary view of fashion to include badges and uniforms.[52] Seen in this light, Darwin's theory of the "taste for the beautiful" on the one hand refers to a philosophical tradition that in itself has a strong biological dimension; this holds particularly for Kant's *Critique of the Power of Judgment*.[53] On the other hand, Darwin's theory is influenced by, and itself part of, a complementary transfer which interprets the clearly cultural phenomenon of fashion—precisely because of its capriciousness—as an analogue of the biological processes of evolution.

Since Darwin's time, numerous studies on animals have convincingly shown that the success of sexual courtship routinely depends on differences of certain features of appearance. The artificial intensification of these characteristics—indeed even the application of additional stimuli (such as attaching little hats to the heads of zebra finches)—significantly improves chances for reproduction.[54] Thus, what Humboldt observed in cultural fashions of human self-ornamentation, becomes in Darwin's model a principle driving the evolution of natural bodies:

---

50 Cf. below, p. 24–27.

51 Darwin, "Development in Dress," 410. Many thanks to Julia Voss for pointing out this text.

52 Spencer, *Ceremonial Institutions*, 174–92.

53 Cf. Kant, *Critique of the Power of Judgment*; Löw, *Philosophie des Lebendigen*; Müller-Sievers, *Self-Generation. Biology, Philosophy, and Literature Around 1800*; Zuckert, *Kant on Beauty and Biology*; Menninghaus, "Ein Gefühl der Beförderung des Lebens."

54 Nancy Burley's experiments with zebra finches are mentioned here as a stand-in for many other studies (see under "Burley" in the bibliography). The experiment with little hats is referred to in Cronin, *The Ant and the Peacock*, 210.

> Not only can we partially understand how it is that man is from various conflicting influences rendered capricious, but that the lower animals are . . . likewise capricious in their affections, aversions, and sense of beauty. There is also reason to suspect that they love novelty, for its own sake.[55]

Darwin's theory of an evolutionarily powerful "taste for the beautiful" stresses the principle of competing for the attention and favor of the audience. In species with high levels of bodily ornament and/or displays of capacities of singing or dancing, the competitive pressure on members of the aesthetically impressive sex (usually the male) is very high. Ornamentation, competitiveness, and combativeness are closely correlating characteristics throughout the animal world.[56] Importantly, Darwin emphasizes that the law of battle is not a substitute for beauty competitions. For a male animal, to win a battle against same-sex competitors does not automatically mean that the female is ready to mate. She must be additionally won over by non-combative devices. This is why Darwin generally ascribes a double function to body ornaments and the arts of singing: they are simultaneously "a challenge to their rivals"[57] *and* a positive means of courting the other sex.

For the nonviolently persuasive effects of aesthetic virtues, Darwin regularly uses the terms "excite," "charm" and "allure," emphasizing their effect upon "emotions."[58] To him, these charms suggests a limitation of the intra-sexual "law of battle": "We clearly see that the plumes and other ornaments of the male must be of the highest importance to him; and we further see that beauty in some cases is even more important than success in battle."[59]

The prevalence of the "taste for the beautiful" over the law of battle encourages a more refined and "more peaceful kind of contest."[60] This hypothesis of a violence-reducing function of aesthetic practices corresponds to classical assumptions of idealist aesthetics. At the same time, it leads Darwin to an interesting comparative observation. Several insects and birds, he observes, are far more involved in peaceful aesthetic competition (as opposed to the law of battle) than "higher" animals.[61] From this

55 Darwin, *Descent of Man*, I 64–65.
56 Darwin, *Descent of Man*, II 54.
57 Darwin, *Descent of Man*, I 56.
58 Darwin, *Descent of Man*, I 56.
59 Darwin, *Descent of Man*, II 98.
60 Darwin, *Descent of Man*, II 313.
61 Darwin, *Descent of Man*, I 418, II 332–33.

perspective, mammals are actually mostly very "brute" animals":[62] "With mammals the male appears to win the female much more through the law of battle than through the display of his charms."[63]

Against this background, humans are among the "more peaceful" mammals, at least with regard to sexual courtship. After all, they even come very close to the outstanding level of bird aesthetics, both with regard to wearing elaborate visual ornaments and performing artistic practices of singing and dancing. Notably, however, aesthetic competition and choice has—in mammals and elsewhere—a violence-restricting function exclusively in regard to the relations *between* the sexes, not in regard to same-sex competition. Therefore, "decoration and pugnacity" can well co-occur, and actually do so routinely.[64]

Accordingly, Darwin does not set fashionable preferences for all kinds of capricious aesthetic displays categorically apart from the dimensions of fitness driving natural selection.[65] It suffices for his model that subtle differences in ornament and pleasing displays are advantageous in sexual choice, provided that natural fitness is otherwise equal. Neo-Darwinian reformulations, on the other hand, attempt to level the difference between sexual and natural selection in a way that in the end reduces sexual selection to yet another type of natural selection. Thus, the costly signal theory[66] tends to interpret the capacity for great aesthetic expenditure—be it in terms of supporting particularly elaborate sexual body ornaments or difficult performances of singing and/or dancing—*in its entirety* as an indicator of "good genes" (albeit with special constraints for humans). Other indicator theories of physical attractiveness have arrived, for different reasons, at similar conclusions regarding aesthetic preferences.[67] However, there is no conclusive evidence so far that *all* sexually selected "caprices" are indicators of fitness rather than of some arbitrary beauty only.[68] At the same time,

---

62 Darwin, *Descent of Man*, II 332.

63 Darwin, *Descent of Man*, II 239.

64 Darwin, *Descent of Man*, II 54.

65 Cf. Menninghaus, *Das Versprechen der Schönheit*, 71–73.

66 Cf. Zahavi, "Mate Selection: A Selection for a Handicap"; Zahavi, "Decorative Patterns and the Evolution of Art"; and Zahavi and Zahavi, *The Handicap Principle: A Missing Piece of Darwin's Puzzle*.

67 Cf. the details on the "naked skin" indicator theory further on, note 102, pp. 21–22.

68 On the theoretical level, John Maynard Smith's arguments against the handicap theory remain relevant (Smith, "Sexual Selection and the Handicap Principle"). Moreover, numerous empirical studies question the proposition that beautiful ornaments are always an elaborate and reliable indicator of greater (genetic) fitness. Cf. Eberhard, *Sexual Selection and Animal Genitalia*; Cronin, *The Ant and the Peacock*; Skamel, "Beauty and Sex Appeal," 173–200; and Miller, "Evolution of Human Music through

Darwin himself cautiously remarked that "in most cases it is scarcely possible to distinguish between the effects of natural and sexual selection."[69]

Darwin's definition of the "taste for the beautiful" is richer in semantic dimensions than it might seem at first glance. It alludes, sometimes just in passing, to numerous elementary characteristics of "beauty" that have been passed down in relative stability from philosophical aesthetics into new attempts at definition. One formula is particularly rich in resonance and implications, informing Hutcheson[70] and virtually all eighteenth-century treatises on aesthetics no less than the foundational treatise on empirical aesthetics, Fechner's *Vorschule der Ästhetik* (*Preschool of Aesthetics*):[71] namely, "uniformity amid variety." It is the license for, and the prevalence of, enormous unreduced variety—of subtle nuances of all sorts—that distinguishes the highly dynamic unity of "sensate" perception and judgment from that aimed at in scientific cognition. According to Baumgarten's critical diagnosis, subsuming concrete individual phenomena under unifying concepts comes at the price of reducing their variety, vividness, and specificity. For these reasons, the term "variety" became a key concept that reflected the innermost stance of aesthetics to do justice to individual sensuous phenomena in all their variety. The rhetorical concepts of *ubertas* (richness, plenty) and *copia* (abundance) were drawn upon as pointing in the very same direction.[72] Darwin adopted this reasoning and used it particularly often in his observations on body ornaments and artful displays. Birdsong repertoires can serve as a perfect example of abundance and plenty in Baumgarten's sense. Numerous birds can sing dozens or even hundreds of complex songs; some master up to 2,000.[73]

In one respect Darwin appears to have underestimated the actual variety of aesthetic competition. He believed that birds and insects compete

---

Sexual Selection," 338–44. A recent book by Richard Prum (*The Evolution of Beauty*) includes a fundamental critique of the Neo-Darwinist tendencies to reduce Darwin's principle of sexual selection to yet another mechanism of natural selection for "good genes."

69 Darwin, *Descent of Man*, I 256. On the theoretical debate regarding natural and sexual selection, cf. also Lande, "Sexual Dimorphism, Sexual Selection, and adaptation in Polygenic Characters"; Grafen, "Sexual Selection Unhandicapped by the Fisher Process"; and Harvey and Bradbury, "Sexual Selection."

70 Cf. Hutcheson, *An Inquiry into the Original of Our Ideas of Beauty and Virtue.*

71 Cf. Fechner, *Vorschule der Ästhetik*, 53–80.

72 Cf. Baumgarten, Ästhetik, particularly § 116–18.

73 Werner, "Too Many Love Songs: Sexual Selection and the Evolution of Communication," 1. Cf. also Slater, "Birdsong Repertoires," 51, and Todd, "Simulating the Evolution of Musical Behavior." Recent studies have shown that sexual songs sometimes impress the wooed females and the male competitors with different characteristics, cf. Leitão, "Are Good Ornaments Bad Armaments?," 161–67, and Rothenberg, *Why Birds Sing*, 69–91.

either in features or looks or in artistic displays, but not in both domains simultaneously.[74] This idea was based on the many birds that show excellent musical abilities but are of an unspectacular appearance. Today, we know that many birds compete in looks, song, and dance displays at the same time. The human being, too, belongs to the species for which aesthetic competition involves multiple dimensions.

Considering that it is inherently tied to a "rich" variety of sensual characteristics, aesthetic perception requires a high tolerance of and pronounced competence in regard to complexity. Experimental aesthetics supports this view.[75] According to classical theory, *ubertas* and *variety* become aesthetically appealing only when they also offer an *order* and *unity* of their own, one that is categorically different from conceptual order. It is this combination of potentially conflicting characteristics that creates the *unlikely* character[76] and the particular appeal of forms experienced as aesthetically pleasing, beautiful, or well-executed. In Darwin's reflections on aesthetics, the aspect of gestalt-like "unity" is far less pronounced than that of exuberant "variety" and ever-changing dynamics. Still, several of his observations are based on both concepts. One example is the distinction he makes between the *calls* and *songs* of birds.[77] Songs tend to be more complex than calls and to contain far more variety. Unlike calls, the individual parts of song sequences have no decodable meaning. And yet, for all their complexity and variety, they still create the impression of an aesthetically convincing unity, for which Darwin uses the expression "to sing their song *round*."[78]

2. Darwin's Visual Aesthetics: From the Theory of Bodily "Ornaments" to the Human Visual Arts

2.1 Naked Skin as the Prime Ornament of Human Appearance

Darwin's reflections on the aesthetic evolution of the human body do not tend to concur with the paradigms of contemporary research on attractiveness. Darwin made no reference to waist-to-hip-ratios, baby face features, body mass indices, or any of the other body features highly prized in

74 Darwin, *Descent of Man*, II 56, 226, 352.

75 Cf. Berlyne, *Aesthetics and Psychobiology*, and Berlyne, *Studies in the New Experimental Aesthetics*.

76 Cf. Gehlen, "Über instinktives Ansprechen auf Wahrnehmungen," 109.

77 Darwin, *Descent of Man*, II 51. For a more extensive treatment of this distinction see below, p. 31–33.

78 Darwin, *Descent of Man*, I 55.

today's cult of beauty. Only one "ornament" interested him with respect to the human body, yet he considered it crucial: naked skin. This "ornament" is usually not even considered to be one. At the same time, the absence of body hair[79] clearly is the most striking difference in appearance between Homo sapiens and other primates:

> The absence of hair on the body is to a certain extent a secondary sexual character; for in all parts of the world women are less hairy than men. Therefore we may reasonably suspect that this is a character that has been gained through sexual selection. We know that the faces of several species of monkeys, and large surfaces at the posterior end of the body in other species, have been denuded of hair; and this we may safely attribute to sexual selection, for these surfaces are not only vividly coloured, but sometimes, as with the male mandrill and female rhesus, much more vividly in the one sex than in the other. . . . So again with many birds the head and neck have been divested of feathers through sexual selection, for the sake of exhibiting the brightly-coloured skin.[80]

For Darwin, naked human skin does not mean absence of clothing and thus the display of genitalia. Rather, the very hairlessness of skin is in itself the first and foremost ornament of the human body. Unlike almost every other ornament considered by Darwin, naked skin is *a full-body ornament*, rather than an addendum at some particular point of the body. Darwin identifies the hairless body parts in the genital area of many monkeys and in the face of the mandrill, which work as sexual signals,[81] as the evolutionary point of departure for the development of naked human skin.[82] Guided over a very long period of time by a preference for regions of hairless skin, sexual selection increasingly amplified this feature of ape attractiveness. Despite practical disadvantages—such as the loss of thermal and mechanical body protection—Darwin credits this process with being powerful enough to bring about near-total denudation,[83] particularly of the female

---

79 Darwin, *Descent of Man*, II 375–82.

80 Darwin, *Descent of Man*, II 376–78.

81 Darwin, *Descent of Man*, II 291–93, 376–78.

82 On the complex relationship between male and female sexual ornaments in primates and on their different functions, see Wickler, "Socio-Sexual Signals and Their Intra-Specific Imitation among Primates," 69–147.

83 In absolute numbers, humans have as much hair on their bodies as many other primates. But this hair does not have the consistency of fur; compared to the hair on fetuses, monkeys, and apes, it is so insignificant as to create the optical impression of hairlessness. Cf. Montagna, "The Evolution of Human Skin," 4–6, and Morris, *The Naked Ape*, 41–42.

body.[84] As a result, a "capricious" taste for areas of naked skin that we essentially share with many apes has been pushed to the point of producing a whole body version of what used to be the sexy hairless spots on the apes' bodies, and by implication an almost entirely sexualized human body surface. Conforming to this intriguing hypothesis, humans tend to treat their entire body surface as erogenous and responsive to stimulation through gentle touch. According to Darwin, we are literally a maximized erogenous zone, a full-body version of the hairless anal-genital area of our ancestors.

The relative hairlessness of the human body is moreover highlighted by the contrast with the head hair, which is particularly rich in women, producing a whole range of complementary hair aesthetics (color, brilliance, texture, fragrance, movement) on an otherwise almost hairless body.[85] The aesthetics of human head hair corresponds to a general focus of ornamentation in many creatures: "The head is the chief seat of decoration."[86] Many other primates, too, boast ornamental hair—"monkey hairdos" or "beards"—around their faces.[87] At the same time, these similarities in ornamental facial hair have a different—if not inverse—frame of reference in that the human body appears naked, whereas the nonhuman primates' bodies are fully covered with hair. And there is yet another "capricious" trait which highlights the near-complete inversion of the distribution of hair and naked skin in humans and their predecessors: pubic hair. In this case, evolution has added hair precisely where the nonhuman primates display their only hairless hotspot (i.e., their genital area).

Darwin nowhere claims that human skin—whether before it lost most hair or after—is not also involved in other biological functions. What he does is try to make a case for the assumption that "to a certain extent"[88] the extreme mutation from an almost completely hairy primate to the almost completely "naked" human is *also* the result of an aesthetic preference driving sexual selection. For him, the denuding of the human being has followed the same

---

84 Darwin, *Descent of Man*, II 291–93. On gender- and population-specific aspects of hair and its absence, see Toerien and Wilkinson, "Gender and Body Hair: Constructing the Feminine Woman"; Tiggemann and Lewis, "Attitudes toward Women's Body Hair: Relationship with Disgust Sensitivity"; Tiggemann and Kenyon, "The Hairlessness Norm: The Removal of Body Hair in Women"; Rebora, "Lucy's Pelt: When We Became Hairless and How We Managed to Survive"; Dixson et al., "Human Physique and Sexual Attractiveness in Men and Women."

85 Cf. Etcoff, *Survival of the Prettiest*, 120–29.

86 Darwin, *Descent of Man*, II 71.

87 Darwin, *Descent of Man*, II 306–13.

88 Darwin, *Descent of Man*, II 376.

self-reinforcing dynamics that can lead a fashionable trend for short skirts to a near-complete exposure of bare legs. If invited to create a new body fashion departing from the monkey look through pushing some originally minimal deviations to an ever higher degree, a top designer could have hardly produced a result as convincing as the process of sexual selection did by implementing a near-complete inversion of how the binary code "hair-covered versus hairless" applies to the body surfaces of apes and humans.

Such purely ornament-driven amplifications of difference and contrast are frequently observed in close relatives: "In many taxa of arthropods and vertebrates closely related species differ most in secondary sexual characters."[89] Darwin's observation—"To our taste, many monkeys are far from beautiful"[90]—thus corresponds to a general rule of "fashion"-based distancing among closely related species. On an evolutionary scale, this phenomenon helps avoid hybrid mating and thus supports the isolation of both a given species and an emerging new one (speciation). Differences in appearance arising from sexual selection are therefore particularly relevant to the differentiation of previously identical organisms into distinct species.[91] In fact, many species of insect can be distinguished solely by their sexual ornaments (including bizarre penis fashions),[92] while a number of bird species can be distinguished only in light of their sexually favored coloring.[93] Body fashion differences of this kind quite literally make many species what they are, particularly among closely related creatures.

Evolutionary theory is thus able to explain why Hesiod chose the monkey as the epitome of ugliness,[94] why Burke stated "there are few animals which seem to have less beauty in the eyes of all mankind [than monkeys],"[95] or why in Goethe's *Elective Affinities* monkeys are portrayed—in a fashion that only seems paradoxical—as both "human-like" and aesthetically "appalling."[96] What is at stake here is the aesthetic rejection of one's own earlier

89 Lande, "Genetic Correlations Between the Sexes in the Evolution of Sexual Dimorphism and Mating Preferences," 84.

90 Darwin, *Descent of Man*, II 310.

91 Cf. Gould and Gould, *Sexual Selection*, 89–90, and Lande, "Models of Speciation by Sexual Selection on Polygynic Traits."

92 Cf. Eberhard, *Sexual Selection and Animal Genitalia*.

93 Darwin, *Descent of Man*, II 230.

94 Hesiod, DK 22 B 83.

95 Burke, *A Philosophical Enquiry*, 105.

96 Goethe, *Elective Affinities*, 137. In his treatment of "Humans' Aesthetic Appreciation of Non-Human Animals," Davies has not included this pronounced aesthetic disliking for our closest evolutionary relatives at least within the Western tradition (*The Artful Species*, 65–85).

body fashion, which is now seen as radically "uncool." In an at least remotely analogous fashion, Walter Benjamin identified one's own parents' choice of clothing as every generation's "most radical anti-aphrodisiac imaginable."[97]

The majority of the hypotheses on human naked skin advocated so far enjoy only limited support in recent discussions.[98] The same applies to the hypothesis that has been dominant for quite some time, namely, that naked skin evolved as a cooling system for long-distance running in the savanna heat. Currently, a hypothesis that had been proposed by Darwin's contemporary Thomas Belt enjoys increased support: naked skin as a protection against ectoparasites.[99] It lies beyond the scope of the present book to discuss the potential compatibility of this hypothesis, or of the many other hypotheses about naked skin, with Darwin's explanation based on aesthetic choice and sexual selection. Suffice it to point out two aspects that can be explained by selection for the capricious beauty of naked skin, but not by the parasite-reduction model:

1. Human evolution has *added* hair, of all places, in an area that is particularly precarious in hygienic terms and that is hairless in many other primates. This addition—pubic hair—is doubtless a sexually selected characteristic: just like other sexual ornaments, it emerges at the time of sexual maturation. Moreover, pubic hair is sexually dimorphous. Its distribution in the genital area has different patterns in males and females. These characteristics flagrantly contradict the hypothesis that hairlessness evolved for hygienic reasons.[100]

---

97 Benjamin, *The Arcades Project*, 64.

98 Cf. Queiroz do Amaral, "Loss of Body Hair, Bipedality and Thermoregulation," 357–66; Rebora, "Lucy's Pelt: When We Became Hairless and How We Managed to Survive"; Rantala, "Evolution of Nakedness in Homo Sapiens"; and Pagel and Bodmer, "A Naked Ape Would Have Fewer Parasites," 117.

99 Cf. Belt, *The Naturalist in Nicaragua*; Rantala, "Human Nakedness: Adaptation against Ectoparasites?"; Pagel and Bodmer, "A Naked Ape Would Have Fewer Parasites"; Rantala, "Evolution of Nakedness in Homo Sapiens," 4 sqq.

100 Pagel and Bodmer are aware of this inconsistency ("A Naked Ape Would Have Fewer Parasites," 118 sqq.). They speculate that in this particular case the selection for parasite-reducing loss of hair was thwarted by another evolutionary aspect, namely, the selection for the transfer of sexual messenger substances (pheromones), to which pubic hair contributes. However, the real challenge for the parasite-reduction hypothesis is not the pronounced exception to the general trend of hairlessness as such but the direct reversal: the transformation of a hairless body part into a hair-covered one. Moreover, humans have so many sweat glands that pubic hair is hardly the only opportunity for pheromone transport. Countless other species manage that function without developing a characteristic with high maladaptive costs.

2. If the human preference for naked skin was primarily based on an evolutionary preference for parasite-free skin, then one might expect humans to find other hairless creatures aesthetically appealing. Naked mole rats have the same advantages as humans regarding protection against ectoparasites,[101] yet they are not considered beautiful. Similarly, hairless cats and dogs can easily be bred, but they are seldom preferred to furry ones. The parasite reduction hypothesis cannot explain the divergent aesthetic evaluation of naked skin in humans and animals. Darwin's model, on the other hand, because it also considers aesthetic preferences as strategies of species separation, can well explain why our preference for naked skin appears to be strictly limited to our own species. Apparently, this is because hairless skin sets us apart from other primates and defines our very special visual identity. For the same reason, we should find any parallel in another species repulsive. (However, this does not apply to aquatic or semiaquatic animals—the latter including walrus, rhinoceros, elephants, etc.—whose environments routinely encourage hair reduction, thus ruling out that hairlessness is a trait distinctive of a particular species.)[102]

---

101 Ibid., 118.

102 Whereas the parasite reduction hypothesis sees the "beautiful" characteristic as immediately and directly adaptive in the sense of natural selection, indicator and signal theories ascribe indirect adaptive qualities to attractive characteristics. The indicator hypothesis states that naked skin which is regarded as beautiful—regardless of whether it is a handicap or not—indicates good health, a good immune system, and thus "good genes" (cf. Hamilton and Zuk, "Heritable True Fitness"; Symons, "Beauty Is in the Adaptations of the Beholder," 96 sqq.; Gangestad and Buss, "Pathogen Prevalence and Human Mate Preferences"; Thornhill and Gangestad, "Human Facial Beauty. Averageness, Symmetry, and Parasite Resistance"; and Wegner et al., "Parasite Selection for Immunogenetic Optimality"). Correspondingly, blemishes and disfigurations of skin have long been regarded as symptoms of diseases. In the framework of the handicap theory (also known as costly signal theory, cf. Zahavi, "Mate Selection: A Selection for a Handicap"; Zahavi, "Decorative Patterns and the Evolution of Art"; Zahavi and Zahavi, *The Handicap Principle*), on the other hand, it is the nonadaptive features of naked skin—its drawbacks in regard to thermal and mechanical protection—that reliably indicate good fitness. The logic here is as follows: one who can afford such a handicap without suffering from it proves otherwise good genes and/or an abundance of resources. Neither of the hypotheses that regard naked skin as a fitness indicator can explain the pronounced human aversion against naked skin in other animal species; nor does this aversion fit in with any hypothesis that suggests direct health benefits. Regardless of their other merits, none of these explanations can replace the explanatory power of Darwin's specific model—though, this hypothesis, too, has its limits.

Usually, sexual ornaments only develop with the beginning of puberty. This is not the case with naked skin—an argument occasionally used against Darwin's interpretation. However, a closer look renders this argument invalid. A fetus is covered in downy fuzz (*lanugo*); after birth, human skin is naked in appearance and gradually goes through a complex redistribution of both hairiness and hairlessness. First, the overall absence is contrasted with the hair on the human scalp; notably, this hair, too, is sexually dimorphous in that it is typically expressed more strongly, and with different ontogenetic changes, in the female compared to the male sex. During sexual maturation, additional hair emerges under the armpits and in the genital area, in the latter case with different distributions for women and men. Moreover, regarding visible hair throughout the rest of the body, the male body shows—even though not in all populations—stronger remainders of nonhuman primate furriness. Thus, replacing the gender-neutral hairlessness of young boys and girls, *a contrast arises* such that female "nakedness" of the bodily parts, like many other sexual ornaments, becomes a sexually distinctive ornament only in puberty.

Summing up, the distribution of naked skin and hairy parts in humans involves a complex set of minor and major contrasts that all share two characteristics: (1) they strongly mark the difference in physical appearance between human and nonhuman primates, and (2) they are sexually dimorphous within the human species and hence play a role in sexual signaling and sexual attractiveness. All these characteristics support Darwin's hypothesis. Classical aesthetics was quite right, then, to defend nakedness as an indispensable aspect of representations of the human body in (Greek) sculpture. Karl Philipp Moritz even highlighted the importance of naked skin for setting humans apart from most animals and hence for their very humanness: "Nakedness, while distorting all other animals, is with the human being the signature that seals and authenticates the perfection of her/his beauty."[103]

Thus, classical aesthetics supports Darwin's notion that we are, as Desmond Morris succinctly put it, *the naked ape*.[104] Naked skin is not just the absence of fur or of garment, but in itself the very natural and highly distinctive "garment" chosen by humans over thousands of generations. It is thus analogous to the most elaborate furs and feathers, even though, to use Quintilian's classification of rhetorical features of speech, it works via *detractio* rather than *adiectio*.[105] At the same time, provocatively proceeding from pink naked monkey behinds and comparing human skin to peacock

103 Moritz, "Die Signatur des Schönen," 582. See also Herder, "Plastik," and "Studien und Entwürfe zur Plastik."

104 Morris, *The Naked Ape*.

105 Quintilian, *Institutio oratorica* IX 3, 18–65.

fans and stag antlers, Darwin's reinterpretation of the classical aesthetics of sculpture also involves an anticlassical turn toward the arabesque and the grotesque. Considered in comparative and functional terms, the delicately curved continuum of human skin and its subtle gradations in color tone—celebrated by classical aesthetics as the natural and discreet decoration of the human body—is in itself a fairly crazy body fashion, fully equivalent to the most elaborate and most colorful ornaments in art and the animal world, or perhaps even surpassing all these.

2.2 The Human Arts of Self-Painting, Self-Decoration, and Self-(De)formation

The first bridge to the visual arts emerges at this point. Most "decorative" arts of Homo sapiens presuppose the evolution of naked skin to achieve their aesthetic effects. The "ornament" of naked skin creates ideal conditions for the artificial application of secondary ornaments. On the one hand, we humans are an antithesis to the aesthetics of peacocks and birds of paradise, as we have placed our bet on the ornamental strategies of detractio—the purist laying bare of body contour—instead of adiectio. On the other hand, we are likewise masters of excessive artificial ornamentation. In the end, it is this superimposition of opposite strategies of ornamentation that defines the aesthetics of human appearance.

"Men of all races," writes Darwin, "take pleasure in . . . painting, tattooing, and otherwise decorating themselves."[106] Self-painting is probably the oldest form of painting. From ochre findings, we can conclude that these practices date back some 150,000 years, perhaps even more than 250,000 years. For at least 80,000 to 120,000 years humans have also produced diverse decorative objects to be worn on the body,[107] in particular chains made from seashells and animal teeth. These objects were perforated, often colored and strung together; however, their three-dimensional form was not itself created through human "art." These practices of self-painting and self-decoration with select objects are considerably older than figurative artworks and fully manmade ornaments, the most ancient available examples of the latter dating back some 40,000 years. It is widely assumed that the more archaic ornamental objects as well as body painting

106 Darwin, *Descent of Man*, I 232.

107 Cf. Conard, "Cultural Evolution in Africa and Eurasia during the Middle and Late Pleistocene." The archeological evidence for self-painting and decorative objects is discussed in more detail below, p. 98–102.

and artificial body (de)formations often signaled social rank and membership in certain groups.[108] This is compatible with the function of sexual attractiveness as assumed by Darwin.

The assumption that individuals wore the most elaborate ornaments and paint patterns in the context of social rites and festive events—just like contemporary individuals, who also tend to dress up for special occasions—is also by no means incompatible with Darwin's hypotheses. After all, on such occasions individuals typically take care to be competitive regarding their personal level of ornamentation. In fact, social occasions involving (ritual) dancing and music are particularly convenient forums for showing oneself and meeting potential partners—and probably have served these functions for a very long time.

Darwin goes to great detail in presenting his readers with the immense variety of artificial self-ornamentations. From the top of the head down to the feet, he describes a colorful range of decorative practices for almost every part of the human body. The loose kaleidoscopic character of this overview is repeatedly interrupted both by anecdotes Darwin's correspondents reported and by comments with which Darwin develops his thread of argument. Here are some longer passages from the beginning of the respective chapter:

> Savages at the present day everywhere deck themselves with plumes, necklaces, armlets, ear-rings, etc. They paint themselves in the most diversified manner. . . . In one part of Africa the eyelids are coloured black; in another the nails are coloured yellow or purple. In many places the hair is dyed of various tints. In different countries the teeth are stained black, red, blue, etc., and in the Malay Archipelago it is thought shameful to have white teeth "like those of a dog." . . . In Africa some of the natives tattoo themselves, but it is a much more common practice to raise protuberances by rubbing salt into incisions made in various parts of the body; and these are considered by the inhabitants of Kordofan and Darfur "to be great personal attractions." In the Arab countries no beauty can be perfect until the cheeks "or temples have been gashed." In South America, as Humboldt remarks, "a mother would be accused of culpable indifference towards her children, if she did not employ artificial means to shape the calf of the leg after the fashion of the country." In

108 Cf. Hovers, "An Early Case of Color Symbolism: Ochre Use by Modern Humans in Qafzeh Cave"; Zilhão et al., "Symbolic Use of Marine Shells and Mineral Pigments by Iberian Neandertals"; Tattersall, "Human Origins: Out of Africa"; Watts, "Ochre in the Middle Stone Age of Southern Africa."

the Old and New Worlds the shape of the skull was formerly modified during infancy in the most extraordinary manner, as is still the case in many places, and such deformities are considered ornamental. For instance, the savages of Colombia deem a much flattened head "an essential point of beauty."

The hair is treated with especial care in various countries; it is allowed to grow to full length, so as to reach to the ground, or is combed into "a compact frizzled mop, which is the Papuan's pride and glory." In northern Africa "a man requires a period of from eight to ten years to perfect his coiffure." With other nations the head is shaved, and in parts of South America and Africa even the eyebrows and eyelashes are eradicated. The natives of the Upper Nile knock out the four front teeth, saying that they do not wish to resemble brutes. Further south, the Batokas knock out only the two upper incisors, which, as Livingstone remarks, gives the face a hideous appearance, owing to the prominence of the lower jaw; but these people think the presence of the incisors most unsightly, and on beholding some Europeans, cried out, "Look at the great teeth!" The chief Sebituani tried in vain to alter this fashion. In various parts of Africa and in the Malay Archipelago the natives file the incisors into points like those of a saw, or pierce them with holes, into which they insert studs. . . .

In all quarters of the world the septum, and more rarely the wings of the nose are pierced; rings, sticks, feathers, and other ornaments being inserted into the holes. The ears are everywhere pierced and similarly ornamented, and with the Botocudos and Lenguas of South America the hole is gradually so much enlarged that the lower edge touches the shoulder. In North and South America and in Africa either the upper or lower lip is pierced; and with the Botocudos the hole in the lower lip is so large that a disc of wood, four inches in diameter, is placed in it. Mantegazza gives a curious account of the shame felt by a South American native, and of the ridicule which he excited, when he sold his tembeta, the large coloured piece of wood which is passed through the hole. In Central Africa the women perforate the lower lip and wear a crystal, which, from the movement of the tongue, has "a wriggling motion, indescribably ludicrous during conversation." . . . Further south with the Makalolo, the upper lip is perforated, and a large metal and bamboo ring, called a pelele, is worn in the hole. "This caused the lip in one case to project two inches beyond the tip of the nose; and when the lady smiled, the contraction of the muscles elevated it over the eyes. "Why do

> the women wear these things?" the venerable chief, Chinsurdi, was asked. Evidently surprised at such a stupid question, he replied, "For beauty!"[109]

And so on, over many pages. In the meantime, archaeology has provided evidence that many of the practices of self-ornamentation included in Darwin's ethnographical panorama are likely to date back to the Middle and Late Paleolithic Eras.[110] Darwin took great care to report local and cultural variants of these practices rather than to focus on a quest for human universals. This is in line with his basic assumption that aesthetic preferences have an inherent tendency to push preferred features to higher levels or even revert them, and that ongoing change and variation can therefore be expected.

The elaborate production and application of paint, the perfecting of a hairdo over many years, the normative removal of teeth, the painful deformation of feet or head—such practices require strong emotional motivation in order to become established as cultural fashions and socially expected stylizations of one's own bodily appearance.[111] Individuals who fail to conform end up in shame and disgrace: shame if they cannot provide the expenditure for themselves; disgrace if they cannot do so for their children.[112] Darwin concludes that the expected reward and/or the sanctions for (non-)conformity must be enormous in order to support the widespread implementation of very costly and often very painful practices of aesthetic self-enhancement: "Hardly any part of the body, which can be unnaturally modified, has escaped. The amount of suffering thus caused must have been extreme, for many of the operations require several years for their completion, so that the idea of their necessity must be imperative."[113]

The *law of beauty* requires dedication, expenditures, and even sacrifices—up to the willingness to mutilate oneself or one's child. In the final analysis, conscious individual motives are of secondary importance for the

---

109 Darwin, *Descent of Man*, II 338–41.

110 Archeological evidence of skull deformations and tooth abrasion is to be found in Romero, "Dental Mutilations, Trephination, and Cranial Deformation"; Borbolla, "Types of Tooth Mutilations Found in Mexico"; Dingwell, *Artificial Cranial Deformation*; Lie, "Dental Decoration in Aboriginal Man"; Low, "Sexual Selection and Human Ornamentation"; Rogers, *Artificial Deformation of the Head*; Coe, "Art: The Replicable Unit." On self-painting and ornaments, see Kölbl and Conard, *Eiszeitschmuck*, the essays on ornaments in the volume *Eiszeit. Kunst und Kultur*, and further archeological references in Chapter 3 of the present book.

111 Cf. Coe, "Art: The Replicable Unit."

112 Darwin, *Descent of Man*, II 340.

113 Darwin, *Descent of Man*, II 342.

establishment of aesthetic self-ornamentations that are perceived as socially imperative. Darwin lists a whole range of possibilities:

> The motives are various; the men paint their bodies to make themselves appear terrible in battle; certain mutilations are connected with religious rites, or they mark the age of puberty, or the rank of the man, or they serve to distinguish the tribes. . . . With the men of New Zealand, a most capable judge (51. Rev. R. Taylor, 'New Zealand and its Inhabitants,' 1855, p. 152.) says, "to have fine tattooed faces was the great ambition of the young, both to render themselves attractive to the ladies, and conspicuous in war." A star tattooed on the forehead and a spot on the chin are thought by the women in one part of Africa to be irresistible attractions.

While religious and social symbols require separate analysis, the functions of sexual attraction and enhancing one's appearance as a fighter correspond exactly to the double message of the birdsong that Darwin describes in a preceding passage.[114]

In the (recent) Western tradition, the arts of self-ornamentation are mostly not included among the arts proper. However, historically and in evolutionary terms, the visual arts of self-painting and self-forming precede painting and sculpture in the narrow sense. The archaic decorative arts imply the following capacities, all of which rely on cultural learning: identification, acquisition, and preparation of suitable painting agents and basic materials (seashells, bones, teeth/ivory, later also metals) for decorative objects; development of instruments and techniques for painting and perforating, in later times also of partitioning and grinding pertinent objects; selection and combination of colors, forms, and materials with the goal of achieving an aesthetically convincing effect. The greatest part of these capacities appears to have been practiced and fine-tuned in (self-) decorative arts for several tens of thousands of years before the first known cave paintings and sculptures were created.

Darwin's reflections on potentially very old practices of visual self-ornamentation are far more elaborate and detailed than his observations on human musical abilities. It is all the more surprising that his theory of the visual arts is almost entirely unrepresented in later research.

---

114 Darwin, *Descent of Man*, II 335–36.

2.3 Seeing the Invisible: From Naked Skin to Aesthetic Imagination

The evolution of naked skin is often considered in conjunction with the human invention of a culture of clothing.[115] Otherwise, it is assumed, we could have hardly survived this "ornament"—or this method of parasite reduction, or of cooling—in most climates. Freud took this assumption as a starting point to propose an interesting addendum to Darwin's theory of naked skin. Combined with cultural clothing, Freud noticed, human nakedness led to a fundamental innovation in the area of sexual fashions: for the first time, parts of a natural body are culturally covered and hence concealed (regardless of how temporary and partial this concealment is). Even if a loincloth is the only item of clothing, naked human skin is divided into *visible* and *invisible* portions. In Freud's perspective, this division entails far-reaching psychosexual consequences:

> The increasing concealment of the body which goes along with civilization keeps sexual curiosity awake which seeks to complete the sexual object through unveiling its concealed parts in imagination. This curiosity can be diverted to artistic efforts ("sublimation") if its interest is turned away from the genitals to the total appearance of the body. The tendency to linger at this intermediary sexual aim of the sexually accentuated gaze is found to a certain degree in most normal individuals; indeed, it gives them the possibility of directing a certain amount of their libido to higher artistic aims.[116]

The distinction between the "visible" and "invisible" creates an entirely new situation for sexual choice dependent on physical attractiveness. Fantasy—the logic of imagination—now fulfills the task of "revealing the sexual object by exposing the hidden parts."[117] This imaginary supplementation has become a crucial part of the erotic interplay between the sexes. More than that: a sparingly or minimally clothed body is often considered more attractive than a completely naked one. In contrast, the hot spots of sexual attraction in all other creatures can invariably be seen (and smelled). In the case of humans, precisely because our skin became naked, plain visibility is

---

115 Cf. Glass, "Evolution of Hairlessness in Man"; Pagel, "A Naked Ape Would Have Fewer Parasites," 5117–19; Kittler, "Molecular Evolution of Pediculus Humanus and the Origin of Clothing," 1414–17; for a different view, see Rantala, "Evolution of Nakedness in Homo Sapiens."

116 Freud, "Three Essays on the Theory of Sexuality," 156–57 (translation by A. A. Brill, with minor changes by A. Berlina and Winfried Menninghaus).

117 Ibid.

doubly limited: on the one hand, clothes render parts of the body invisible; on the other, imagination fills the lacuna of invisibility through its own anticipations and phantasies (which can arouse sexual and other interest in their own right).

For Freud, this is just as important an aspect of evolution as the "denudation"[118] of skin. Its consequence is that the "beautiful" and "stimulating" parts of a sexual body are partly shifted into the sphere of imagination. This substantially contributes to the possibility that aesthetic attraction is diverted from the immediate pursuit of sexual goals. This "diversion" is conceived not as primarily based in prohibition but as a special effect of the human cultural development toward combining a biologically evolved and highly ornamental nakedness with an equally ornamental concealment of this cardinal ornament. If imagination plays a role already in mere acts of looking at "sexual objects," then it does not have to intervene from a reality-transcendent position in order to reorient sexual arousal toward "higher" imaginary goals. If so, this culture-creating power par excellence is involved, in an originary fashion, in the very mechanisms of sexual attraction itself. And if so, the human choice of sexual partners based on appearance differs from that of all other species not only gradually but categorically: it is tied to an unprecedented *becoming imaginary of the body image* that is co-emergent with concealing clothing.

Only in this field could bodily *beauty* acquire dimensions of a *hidden secret*, and theorems of *veil* and *cover* could become ingredients of the very concept of beauty.[119] Freud's interpretation of Darwin's theory thus provides novel perspectives on the desiderata of traditional philosophical aesthetics. The latency-driven intervention of imagination into the field of sexual choice based on physical attractiveness—which for Darwin entailed no visual latency, but only patent ornaments—revolutionizes the entire correlation of aesthetic evaluation and sexual desire. It transforms the motivational component of aesthetic evaluation no less than the field of visibility: henceforth, the orientation toward a sexual goal is partially diverted and turned into a motor of imagination.

Summing up: the specific human development toward a new kind of beauty—and toward naked skin in particular—opens up a new *in*visibility; this intervention of invisibility into the field of looks-based evaluation triggers imaginative processes of supplementing the missing visual

118 Darwin, *Descent of Man*, II 378.

119 Cf. Walter Benjamin, "Goethe's Elective Affinities," 333–56.

information; together, these two innovations transform aesthetic appreciation into a tense field in which archaic sexual and cultural mechanisms interact.[120]

Darwin noticed that the curious human fashion of naked skin was an ideal prerequisite for the cultural arts of visual self-ornamentation. Freud expanded this insight by drawing an additional conclusion for the field of human aesthetics. Naked skin by itself supports a cultural deflection of the very power (namely, sexual choice) that has led to its emergence. Thanks to its biological features, naked skin encourages the development of the cultural arts of ornamentation, concealment, and supplementation. The evolution of naked skin thus prepared the stage for multiple configurations of patency and latency, showing and concealing, secret/mystery and revelation, veiling and unveiling—configurations that are key to human aesthetics.

3. Darwin's Theory of Music and Rhetoric

3.1 Birds, Mammals, and Humans as Sexual "Singers"

The plumage of the peacock and the great argus pheasant are emblematic examples of Darwin's theory of beauty-driven sexual selection. Moreover, birds in general are for Darwin "the most aesthetic of all animals."[121] They excel in all areas of aesthetic self-presentation: as bearers of lavish ornaments, artful dancers, masterful builders, and singers. For ages, humans have held the aesthetic splendor of many birds in high esteem, exploiting them for their own purposes. The widespread human practice of wearing feathers as decorations adopts avian ornaments for human purposes, thereby enriching human aesthetic potency.[122] The metaphor "to adorn oneself with borrowed plumes" reflects both these practices and the general importance of the bird model of competitive displays of ornaments far beyond Darwin's account. Peacocks have played a prominent role in the cultural history of human ornithological self-adornment.[123] Similarly, the singing of birds has again and again been regarded as a model for human musical endeavors, and has also been drawn on by musicians.[124]

120 Cf. Benthien, *Haut*.

121 Darwin, *Descent of Man*, II 39.

122 Darwin, *Descent of Man*, II 372.

123 Cf. Reimbold, *Der Pfau: Mythologie und Symbolik*.

124 Cf. Rothenberg, *Why Birds Sing*, 17–29.

In their ability to articulate distinct vocal sounds, let alone words, the great apes are vastly inferior to many other animals such as parrots or elephant seals. Still more modest are their capabilities to sing melodically or learn a rhythm. Accordingly, the vocal tract of apes features several strong differences from a human one (position of the larynx, relation of tongue and lips, flexibility of the oral cavity for sound production, etc.). Moreover, the great apes do not possess the neural and motor mechanisms needed for finer vocal articulation or even artful singing.[125]

Songbirds, by contrast, masterfully command such mechanisms, as do humans. The similarity of some physiological preconditions for "music" in humans and songbirds was known already to Darwin. Other similarities—in brain construction, neural processing, and individual genetic preconditions—have been discovered only recently.[126] Most elaborate bird "songs" are produced only during mating seasons, and the musical abilities and articulatory organs are typically unevenly expressed in male and female birds. Sometimes only the courting sex (mostly the male), but not the wooed one, has such organs. In cases where both sexes have the same articulatory capabilities, the male sex hormone has been shown to cause differences both in their expression and in the neural correlates of singing.[127] Many insects, amphibians, and aquatic mammals also dispose of high articulatory capabilities and typically also use them in the context of sexual courtship.

However, precisely (terrestrial) mammals offer relatively little evidence for active musical abilities and practices. To be sure, many mammals, too, show sexually dimorphous characteristics of the vocal organs, with males making particularly frequent use of these organs during sexual display. But they do not sing. As a rule, they do not produce anything coming close to the rhythmic and/or melodic displays observed in the countless mating songs of other animals. Darwin categorically distinguished between "calls" and "songs."[128] Calls vary only very little within a given species; they are short, structurally simple, and readily decodable regarding their meaning. They signal an affective response to a given situation, such as readiness for aggression or submission, as well as alarm, distress, or something similar. They often serve to prime other individuals toward particular action responses. Some alarm calls of birds, nonhuman primates, and other

125 Jürgens, "Neural Pathways Underlying Vocal Control," 235–58.

126 Cf. Falk, "Hominid Brain Evolution and the Origins of Music," and Haesler and Scharff, "Genes for Tuning Up the Vocal Brain: FoxP2 in Human Speech and Birdsong."

127 Cf. the summary of the corresponding studies in Rothenberg, *Why Birds Sing*, 154.

128 Darwin, *Descent of Man*, II 51–52.

animals not only trigger an alarm in general but also indicate the class of potential attackers, thus involving functional characteristics of referential symbolization.[129]

Songs, on the other hand, are rich in variation, in many cases dependent on long practice,[130] usually more temporally extended than calls, and syntactically as well as melodically complex. Last but not least, songs typically do not allow, and do not call for, a referential (symbolic) or emotional decoding of the meaning of their individual musical sequences.[131] Clearly, these highly structured tone sequences have a sexually courting function and hence aim at winning over individuals of the opposite sex; and yet, this affect-regulatory performance can hardly be broken down into referential signals sent by the individual musical subunits. Rather, it is the various parameters of *the elaboration of the signal itself* that "impress" the courted or competing individuals. Some authors hold that sexual selection in favor of ever more complex song sequences took the far simpler species-specific calls as points of departure.[132]

Darwin fully acknowledged the multiple nonsexual functions of vocal calls.[133] For the comparative definition of human singing, however, he primarily sought examples for the special type of vocal behavior he called the "true song."[134] Observations of birdsong also underlie Darwin's criterion that, in order for a vocal display to be eligible to be called a "true song," it must go through an often very time-consuming process of exercise and perfection through learning.[135] Against this gold standard, Darwin arrives at the sobering conclusion that there are barely any mammals practicing genuine "song" for the purpose of sexual courtship.[136]

There are many reasons why natural selection would not support elaborate singing. By its definition, "true song" is metabolically expensive: the physiological preconditions for singing must be developed and supported, the singing ability must be trained and refined, and the singing itself often takes up much time and energy in the process of courting.[137] Moreover,

129 Cf. Cheney and Seyfarth, *How Monkeys See the World.*

130 Darwin, *Descent of Man*, I 55–56.

131 Cf. Hockett and Ascher, "The Human Revolution"; Fitch, *The Evolution of Language*, 470–72; and Rothenberg, *Why Birds Sing*, 6–7, 65–68, 72–79.

132 Cf. Marler, "Origins of Music and Speech," 41; and Geissmann, "Gibbon Songs and Human Music," 118.

133 Darwin, *Descent of Man*, II 51.

134 Ibid.

135 Darwin, *Descent of Man*, I 55–56.

136 Darwin, *Descent of Man*, II 332.

137 Cf. Zahavi, *The Handicap Principle*, 28.

loud singing gives away the singer's presence and location, increasing the risk of being attacked by predators. Most songbirds can fly very well and are not an easy catch for larger animals. Thus, it seems that the circumstances of their ecological niche have not led to sufficiently strong selection pressures against singing. Revealing their coordinates to predators may be more problematic for mammals than for songbirds. However, even in the absence of genuine singing many mammals do give away their location through heightened vocal activities during the period of sexual courtship. Therefore, this factor alone cannot explain the absence of singing.

Exceptions to the rule that mammals are not disposed to elaborate, partly learned, and variable singing do exist: humpback whales, for instance, sing very complex songs during the mating season. Darwin supports his views on the human ability to sing by referring to two singing primate species: howler monkeys and gibbons.[138] A recent survey suggests that twenty-six primate species (meaning as many as 11 percent of all primates) produce complex and temporally extended sound sequences.[139] Moreover, the loud calls of our closest relatives, chimpanzees, even share several acoustic parameters with the song of gibbons. At the same time, these sounds are even less eligible to be called "singing" than those produced by gibbons. Today, three functions are ascribed to the "singing" of primates: social coordination of groups, partner behavior (duets), and sexual courtship according to Darwin's model. These functions appear to be important to a different extent to different species; in several species, two or all three functions are assumed to be served by such vocalizations.

However, all these data are subject to an important limitation: if we side with Darwin in regarding vocal learning and the concomitant increase in ranges for (creative) variation as a necessary requirement of "true song," then the human being is *the only singing primate*![140] Unlike with birds, whales, or elephant seals, who do exhibit vocal learning, the repertoire of nonhuman primates seems to be entirely innate.[141] At the same time, there are birds, too, whose rather complex songs appear to be inherited to a large degree. It may seem counterintuitive at first sight to make the categorization of vocal skills so strongly dependent on the way they are acquired. However, when it comes to a comparative evolutionary view of the *arts* of singing, the capacity for learning, perfecting, and refining is an indisputably important factor. The ability for vocal learning evolved independently in

---

138 Darwin, *Descent of Man*, II 332.

139 Geissmann, "Gibbon Songs and Human Music," 112.

140 Fitch draws this consequence; see "The Biology and Evolution of Music," 11.

141 Geissmann, "Gibbon Songs and Human Music from an Evolutionary Perspective," 108.

unrelated species; still, it is quite rare overall. Among mammals, it is found far less frequently than among birds, and, again, among all primates *only humans* seem to have developed this ability.[142] (Analogously, with dance skills, the complexity of a relatively long sequence is not yet sufficient for an analogy with the greatly variable human dance arts. The courting dance of sticklebacks, for example, is immensely complex, but always follows the very same apparently innate pattern.)

At least three factors are prerequisites for the capacity for singing in Darwin's sense: (1) the development of a suitable vocal tract, (2) the ability for neural fine-tuning of the vocal tract, and (3) the capacity for vocal learning. Some singing animals—such as crickets, frogs, some birds, and all singing primates—have only developed features (1) and (2). The great apes have not adequately developed any of the three prerequisites for singing. Their vocal tract, similar as it is to that of humans, lacks some essential modifications. Its neural control is very poorly developed, and there appears to be no vocal learning whatsoever. Humans, on the other hand, score very high on all three capacities, just like many songbirds do. This gap between human and nonhuman primates—and hence the absence of evidence for a gradual evolution—underscores the magnitude of the problem Darwin faced in explaining human musical abilities.

This problem is somewhat alleviated by the fact that the great apes developed the instrumental "music" of drumming—a stable characteristic of human music, too, according to the findings of ethnomusicology.[143] Whereas gorillas mainly drum on their own bodies, chimpanzees and bonobos also use tree roots and other objects for the purposeful production of sounds. The drum sequences of chimpanzees, however, hardly exceed the length of a second; in the case of bonobos, up to twelve seconds have been observed.[144] In any event, the gap in the vocal arts of singing is not diminished by this phenomenon, fascinating as it is in evolutionary terms. No signs of "instrumental learning" have been found among apes so far, either.

Darwin acknowledged that he knew close to nothing about the alleged or real "singing" of nonhuman primates.[145] However, reports of gibbons and howler monkeys encouraged him to speculate that there might have been more singing mammals and primates in the course of evolution. The

142 Cf. Janik, "Vocal Learning in Mammals," 10–13, 16.

143 Cf. Pika, Liebal, and Tomasello, "Gestural Communication in Young Gorillas"; Arcadi and Boesch, "Buttress Drumming by Wild Chimpanzees"; Kirschner and Tomasello, "Joint Music Making," 362.

144 Cf. Fitch, "The Biology and Evolution of Music," 22–23.

145 Darwin, *Descent of Man*, II 332.

"half-human progenitor" of modern Homo sapiens could have been such a being, a primate/hominid with the capacity for singing,[146] or at least a being that has already acquired, for purposes other than singing, the vocal abilities that would allow us humans to sing.[147] Through this move, Darwin envisaged a possible evolutionary trajectory of how the musical capabilities of Homo sapiens sapiens could have evolved so dramatically past the musical weakness of most related mammals, including the great apes, without this evolution necessarily being confined to the evolutionarily very short lifetime of Homo sapiens only. Alleviating this problem from another vantage point, recent findings suggest that—despite the large differences mentioned—there are also evolutionarily relevant matches between human and nonhuman primates in their vocal forms of communication.[148]

Darwin did not have access to remotely as much information as we have today regarding the history of human evolution and the sequence of hominids before Homo sapiens sapiens. Therefore, he could not but leave open the question at what point in this millions-of-years-long lineage he expected the "half-human progenitor" to have evolved "musical abilities." In today's evolutionary biology and archeology, several positions are represented that are, in principle, compatible with Darwin's hypothesis:

1. Based on multiple indicators (lower larynx, thereby increased resonance space in the mouth and throat, hypoglossal canal, breath control, brain volume), Robin Dunbar estimates that the human capacity for highly articulate speech dates back between 500,000 and 1.6 million years. In this view, the common ancestors of Neanderthals and anatomically modern humans were already able to produce articulate speech.[149] Some authors argue that Neanderthals might have had syntactic language and music.[150] Others ascribe syntactic and symbolic language exclusively to anatomically

146 Darwin, *Descent of Man*, II 334.

147 Darwin, *Descent of Man*, II 335.

148 Cf. Richman, "Rhythm and Melody in Gelada Vocal Exchanges"; Hauser, "The Sound and the Fury," 98; Geissmann, "Gibbon Songs and Human Music," 118.

149 Dunbar, *The Human Story*, 123–26.

150 Cf. Hagen, "Did Neanderthals and Other Early Humans Sing?," 291–320; and Schenk, "Vom aufrechten Gang zur Kunst," 58. Steven Mithen's book *The Singing Neanderthals* is wholly based on Darwin's key assumption. However, Darwin's rich further development of this idea—for instance, his hypotheses regarding the range and the modality of music-induced emotions—is not considered. See the review by Fitch, "Dancing to Darwin's Tune," 288.

modern humans (200,000 years ago), and attribute to Neanderthals only a-grammatical holistic expressions associated with gestures and musical prosody.[151] If the hypothesis of "singing Neanderthals" should turn out to be well-founded, it is not a far-flung idea that the common ancestor of modern humans and Neanderthals—*Homo heidelbergensis*, now also called "archaic Homo sapiens"—could have had "musical abilities." This archaic Homo sapiens would then be a singing ancestor in Darwin's sense.

2. Darwin's hypothesis that music predates language often finds support in these contexts. According to one evolutionary scenario, rhythmical-musical sounds could first have evolved in the context of multimodal communication (combinations of gestures, body movements, and sounds), as can be found in the great apes. Dunbar suspects, quite in Darwin's vein, that "music has very ancient origins, long predating the evolution of language." Accordingly, the first language may have been "musical rather than verbal."[152] Speculative authors attribute to the archaic Homo sapiens (Homo heidelbergensis) the capacity for full-blown collective singing, about 350,000 years before our time. [153]
3. The music-before-speech hypothesis finds particularly strong support in developmental psychology.[154] Across cultures, mother-to-child communication has a more or less musical patterning, with an over-expression of rhythm and of prosodic characteristics such as a strong role of pitch and speech melody. These features serve to raise and maintain attention, to give affective signals with regard to the given situation, and to maintain an ongoing vocal-emotional bond between the communication partners (affective bonding). Moreover, these proto-musical features apparently make it easier for a child to detect the boundaries of syntactic and semantic units within a given message, and thus promote the learning of spoken language. Some authors interpret this ontogenetic finding as indicating that our capacity to be affected by music has a phylogenetic origin in

151 See the summary of corresponding studies in Mithen, *The Singing Neanderthals*, 226–31.

152 Dunbar, *The Human Story*, 126 and 132.

153 Mithen, *The Singing Neanderthals*, 217–20.

154 Cf. Bråten, *Intersubjective Communication and Emotion in Early Ontogeny*; and Dissanayake, "Antecedents of the Temporal Arts in Early Mother-Infant Interaction."

this special type of mother-child communication which has come to be called motherese or infant-directed speech (IDS), as well as in the practice of singing lullabies, which also seems to share some universal formal features across cultures.[155] Other authors hold that the prelinguistic infant-parent communication might be a major resource not just for music, but also for later interest in poetry and musical elaborations of language.[156]

4. Recent research suggests some universal characteristics of human musical perception. This perception is not only generally sensitive to (relative) pitch, rhythm, and melodic contours; it also seems to prefer tonal systems that share certain basic traits.[157] The human sensorium may also well be transculturally calibrated for some universal, affect-encoding musical features.[158] Moreover, certain patterns of vocal affective expression that have analogues in music are shared with other species.[159] If these hypotheses were confirmed and withstood critical objections,[160] there would be good evidence that we produce and appreciate music on a special evolved basis dating back deep into our evolutionary past. On their own, however, such musical universals would not per se support the functional hypothesis of a sexual selection–driven origin of music, or any other functional hypothesis.

Importantly, Darwin did not claim that a selection pressure for singing in sexual courtship must have been the causal driving force underlying the evolution of human vocal abilities. On the contrary, he expressly admitted the possibility that these abilities might have evolved for other purposes:

155 Cf. Falk, "Prelinguistic Evolution in Early Hominids: Whence Motherese?"; Unyk et al., "Lullabies and Simplicity"; Trehub et al., "Parents' Sung Performances for Infants"; Trehub et al., "Infants' and Adults' Perception of Scale Structure"; and McDermott and Hauser, "The Origins of Music," 33–34.

156 Cf. Dissanayake, "Antecedents of the Temporal Arts in Early Mother-Infant Interaction." For similar assumptions held in German Romanticism see Kittler, *Discourse Networks 1800/1900*.

157 Cf. the study report on "Universal Features of Music" and "Innate Sensitivity to Musical Structure," in McDermott and Hauser, "The Origins of Music."

158 Cf. Fritz et al., "Universal Recognition of Three Basic Emotions in Music," 573–76.

159 Cf. Juslin, "Communication of Emotions," 773; Leinonen, "Vocal Communication between Species: Man and Macaque," 241–62; Leinonen, "Shared Means and Meanings in Vocal Expression of Man and Macaque," 53–61.

160 Cf. Justus and Hutsler, "Fundamental Issues in the Evolutionary Psychology of Music."

> Many cases could be advanced of organs and instincts originally adapted for one purpose, having been utilized for some distinct purpose. Hence the capacity for high musical development which the savage races of man possess, may be due either to the practice by our semi-human progenitors of some rude form of music, or simply to their having acquired the proper vocal organs for a different purpose.[161]

In the rendering of Darwin's hypothesis that is customary today, this sentence and this distinction usually make no appearance. A simplified "singing for sex" hypothesis is widely understood as a hypothesis about the evolution of our vocal abilities. Darwin, however, carefully avoided any needless commitment to such a narrow view. To be sure, after acknowledging that vocal abilities have "many purposes" for "animals of all kinds," he right away added: "the vocal organs were *primarily* used and perfected in relation to the propagation of the species."[162] But this still leaves open the possibility that the development of vocal abilities could serve more than *one* function. And with regard to humans, the hypothesis of the primacy of sexual selection is subject to an unambiguous limitation: the evolutionary primacy of other purposes is expressly declared possible.

More than that, Darwin even opens his remarks on human vocal and musical capacities with a sentence that virtually denies their sexual origin: "the capacity and love for singing or music, though *not* a sexual character in man, must not here be passed over."[163] Thus, Darwin acknowledged two things right from the start: (1) In contrast to many animals, musical capacities in humans are not subject to a markedly asymmetric distribution between the sexes, as is typical of more clearly sexually selected traits. (2) Sexual selection is no longer the dominant context for the use of these abilities—and perhaps it never has been. It almost seems as if Darwin had trouble explaining why he treats human musical abilities under the general title of *Selection in Relation to Sex* in the first place.

Darwin's reader is not immediately provided with a reason why the human "capacity and love for singing or music . . . must not here [i.e., in the context of sexual selection] be passed over." The formula "must not be passed over" is not exactly a strong argument. Thus, it becomes evident that Darwin starts building his line of argument while fully acknowledging that a "singing for sex" hypothesis is very difficult to maintain for humans. His

161 Darwin, *Descent of Man*, II 335.
162 Darwin, *Descent of Man*, II 330.
163 Ibid., my italics.

correspondence partners from around the world were apparently unable to report any evidence for strikingly vocal or dance-based sexual courtship practices in either nonhuman primates or humans, and neither could he provide such evidence from other sources or personal experience. He also refrained from (ab)using as anecdotal evidence clichéd opera "folklore" in which singing men court ladies standing on Italian balconies. And the sexual success of modern pop stars referred to by Miller[164] was certainly no argument for Darwin on an evolutionary scale.

Still, Darwin did discuss the relationship of human music and sexual selection in the course of the chapter, and particularly toward its end. This relationship exclusively bears on the "capacity for *high* musical development," and not necessarily on the evolution of human vocal skills altogether. Darwin's hypothesis is therefore not per se incompatible with hypotheses according to which the human vocal abilities primarily evolved in mother-child communication, the vocal synchronization of groups, or in response to other natural selection–driven pressures in favor of refined capacities of acoustic communication. Under this premise, it would then only be the *competitive elaboration* of skills already developed "for a different purpose" that Darwin attributes to the "aesthetic" dynamics of sexual selection. Numerous abilities developed for purposes other than aesthetic competition can become "arts" through high degrees of elaboration, whether they lose their old function in favor of a new one or acquire a new function without sacrificing the old one. Accordingly, Ellen Dissanayake holds that all arts are fundamentally nothing but the "making special" or "artification" of everyday practical activities.[165] Darwin's theory of "the capacity for *high* musical development" is compatible with this view.

Just as he did with features of physical attractiveness[166] and the use of decorative ornaments, Darwin sees nearly equal, though differently expressed, singing capacities in women and men, and hence no strong sexual dimorphism as in many birds. The stronger male voice corresponds to the pattern exhibited by most other animals whose males have the courting role.[167] The greater "sweetness" of the female voice, on the other hand, could be read as suggesting "that women first acquired the musical powers in

164 Miller, *The Mating Mind*, 73 sqq.; Miller, "Evolution of Human Music through Sexual Selection," 331. Steinig in his *Als die Wörter tanzen lernten* presents a similar but far more speculative interpretation of the human dance capacity as an evolved "costly signal" of sexual courting. In his reading, human speech is primarily derived from dancing rather than music.

165 Cf. Dissanayake, *What Is Art for?*, *Homo Aestheticus*, "The Arts after Darwin," and Dutton, *The Art Instinct*, 191 sqq.

166 Darwin, *Descent of Man*, II 338–82.

167 Darwin, *Descent of Man*, II 330, 337.

order to attract the other sex."[168] In the end, Darwin appears to favor the hypothesis of "a mutual courtship and rivalry" in which both sexes are equally engaged by way of producing musical tone sequences, "which, at an extremely remote period, . . . aroused each other's ardent passions."

This hypothesis is by no means incompatible with the model of sexual selection. Asymmetry is the norm for sexually selected characteristics, but certainly not the only possible scenario. The fact that Homo sapiens sapiens develops the capacity for singing before puberty—unlike other sexual ornaments—does not per se invalidate Darwin's hypothesis, either. Possibly, the sex-dependent characteristics of the voice—that do correlate with sexual maturity—are the basis for a novel, sufficiently distinctive, and functionally specialized use of the preexisting general capacities for singing. Moreover, there are indications that general vocal skills appear to become more elaborate at the onset of puberty.

In any case, Darwin's standard model—male birds court females, who select their sexual partner according to the quality of his song—is not applicable to the human situation without great reservations. Actually, it only applies with significant constraints even to birds.[169] According to current knowledge, many more female birds engage in singing than Darwin was aware of. In particular, the phenomenon is widespread among previously little-researched tropical birds. Some female birds that usually do not sing nevertheless have the capacity to do so, and can be induced to demonstrate this when stimulated with male sexual hormones. Still, two sex-related asymmetries regarding birdsong appear to be firmly in place. First, highly developed male vocal arts often go along with relative absence of vocalizations on the part of the females; inversely, high female vocalizing does not predict a muteness of male birds. Secondly, there appears to be no bird species in which the females court the males by means of singing and hence in which male sexual choice is based on female song performance. These two observations suggest that female vocalizations have primarily nonsexual functions in birds. Thus, despite many new findings, the model of sexual birdsong remains firmly associated with the asymmetry between male song and female judgment/choice.

---

168 Darwin, *Descent of Man*, II 337.

169 Regarding the evidence referred to in this paragraph, cf. Langmore, "Female Song Attracts Males"; Langmore, "Why Female Birds Sing"; Riebel, "The 'Mute' Sex Revisited"; Ritchison, "The Singing Behavior of Female Northern Cardinals"; Fitch, "The Biology and Evolution of Music," 14 sqq.; Darwin, *Descent of Man*, I 276; Owens, "Extraordinary Sex Roles in the Eurasian Dotterel: Female Mating Arenas, Female-Female Competition, and Female Mate Choice"; Petrie, "Mating Decisions by Female Common Moorhens (Gallinula choropus)," 947–55.

Human cultures known to Darwin do *not* provide sufficiently widespread and concrete examples for birdlike musical displays in the context of sexual courtship. However, Darwin points out: "Love is still the commonest theme of our songs."[170] At least in regard to preserved Western evidence, this finding is difficult to rebut.[171] From Archilochos and Sappho to today's pop songs, occidental song culture can indeed hardly be imagined without the theme of love. Certainly, the ritual singing of songs to the wooed one—let alone large-scale singing competitions involving many participants—is not an indispensable ingredient of human sexual courtship efforts. Even so, love songs are of great emotional value to individuals and societies. Singing such songs can be emotionally meaningful, even if the desired being is not addressed directly and the singer sings only to him- or herself. Similarly, listening to love songs does not necessarily elicit sexual feelings for the singer; it can also awaken or strengthen latent or patent feelings for another person. This associative transfer enabled by imagination provides channels of conjuring up, intensifying, and even anticipating the wished-for realization of romantic and sexual feelings; the object of these feelings does not even need to know about them. Such symbolic practices can well have direct practical functions for the emotional and sexual lives of individuals. They support processes of sexual courtship that tend to be much longer and more complex in humans than in other animals, and that can include prolonged periods of hope and anxious uncertainty.

The processing of disappointments is an important part of what love songs can be good for. Communicating unrequited love in a song—or simply listening to such songs—may (also) be a means of coping with this painful predicament.[172] Such sharing of a real or expected rejection may help fight despair and increase the chances that one does not give up but keeps trying. Thus, the human love song may indeed continue much older practices of sexual courtship in multiple ways. However, it does so within the very special framework of human symbolic communication, and hence can exert emotion-regulatory effects even in the absence of a desired sexual object.

---

170 Darwin, *Descent of Man*, II 336.

171 Spencer ("The Origin of Music," 455–57) pointed at numerous non-Western cultures in which love songs appear to be an exception. However, even if Spencer's observation were correct, it would not by itself refute Darwin's hypothesis. After all, this hypothesis does not necessarily imply that remnants of archaic courtship practices must express themselves in songs whose content is explicitly love-driven, be it exclusively or at least primarily. As shall be explained in more detail, Darwin assigns the relevant affective power of song far less to the underlying text per se (the lyrics), but to melody, rhythm, and emotional prosody.

172 Cf. Rimé, "Emotion Elicits the Social Sharing of Emotion."

Another, more general, parallel between sexual courtship and artistic performance extends far beyond the special genre of love song. Both sexual courtship and artistic performance compete for the limited resource "attention" using supernormal stimuli, and both strive for affection, for "love" or "desire" on the part of the addressees. In the non-performing arts, this desire for the desire of others is less direct but in general no less virulent.

Summing up, none of the arguments Darwin provides in favor of a sexual explanation of human singing is empirically or logically conclusive, let alone apodictically true. Still, all arguments together represent no more—but no less—than indications that the hypothesis may well be worth further consideration. Importantly, Darwin himself emphasized that his notion of a proto-music that may have played an important role in sexual choice must be categorically distinguished from more recent musical practices, and hence is a primarily theoretical construct.

3.2 Darwin's Theory of Music-Elicited Emotions

Partly bridging the gap between a hypothetical proto-music and historically known music, Darwin's theory of human music is primarily an evolutionary theory of its emotional effects. In a first step, Darwin sketches out the range of discrete emotions elicited by "musical tones," "rhythms," and "cadences." Secondly, he emphasizes a specific modality of these emotions.

> Music arouses in us various emotions, but not the more terrible ones of horror, fear, rage, etc. It awakens the gentler feelings of tenderness and love, which readily pass into devotion. . . . It likewise stirs up in us the sense of triumph and the glorious ardour for war. These powerful and mingled feelings may well give rise to the sense of sublimity.[173]

Regarding this panorama, Darwin's quest is that of an archeologist of hypothetically forgotten and only subliminally felt origins. Pursuing this goal, he maps the emotions hypothetically elicited by music onto the scenario of sexual courtship. Efforts of sexual courtship include the two emotional poles he identifies in music. On the one hand, they display tender dedication and "devotion" combined with ardent desire, expectation of and joy/excitement about reciprocity, fear of and sadness about rejection, and are

173 Darwin, *Descent of Man*, II 335.

in any event high in arousal ("ardent passions"). On the other hand, the situation of sexual courtship also involves fierce intra-sexual competition and fights: "The season of love is that of battle."[174] After listing numerous examples from the animal kingdom, Darwin's remarks regarding humans are discreetly limited to "ancient times" and "barbaric tribes":

> Women are the constant cause of war both between the individuals of the same tribe and between distinct tribes. So no doubt it was in ancient times: "nam fuit ante Helenam mulier taeterrima belli causa" [Already before Helen, women were the commonest cause for war].[175]

If this holds true, then sexual courtship can also awaken feelings that are the very opposite of "tender emotions." The term "battle" is by no means overstated in this context. The captivation and abduction of females is well-known as a key cause for violent, warlike actions, be it among chimpanzees or humans.[176] In species with territorial behavior, the male capacity to secure and protect a territory—if necessary, with violent fights—is a decisive factor of female sexual choice. The sexually courting song behavior of many birds simultaneously serves, to different degrees, three functions: attracting sexual partners, discouraging or confronting competitors, and claiming territorial ownership along with signaling readiness for defense—the two latter functions being often, but not always, the same.[177] Darwin's laconic formula regarding love and war encompasses this whole range.

A closer look at this range shows how easily individual emotions of opposite valence can become closely associated. Tender feelings can lead up to loving self-abandonment, to self-sacrificing adoration of the object of desire. Such devotion approaches the worship of a higher being, and thus borders on feelings of the (religious) sublime, or awe. Feelings of readiness for combat and of triumph can likewise be of a grand or sublime nature, albeit differently colored. The range of music-elicited emotions identified in a more recent empirical survey shows at least some overlap with Darwin's daring outline.[178]

---

174 Darwin, *Descent of Man*, II 48.

175 Darwin, *Descent of Man*, II 323. The quotation from Horace, *Sermones* I.3, 107 sqq., reads correctly: "nam fuit ante Helenam cunnus taeterrima belli causa."

176 Cf. Wrangham and Peterson, *Demonic Males*; and Gat, "Why War? Motivations for Fighting in the Human State of Nature."

177 Darwin, *Descent of Man*, I 56.

178 Zentner et al., "Emotions Evoked by the Sound of Music."

Darwin's remark concerning the war about Helen distinguishes two types of sexual competition: "between the individuals of a tribe and between different tribes." The first type only loosely correlates with territorial behavior, the second—which only exists in species that live in social groups—shows a clear correlation. Humans are among the species in which both types of sexual competition are found—all the more so as the rule of exogamy often transcends the confines of one's own group. Music specifically practiced during wartime belongs to the best-documented transcultural variants of human music.[179] Chimpanzees also increase their preparedness for aggression against neighboring populations by means of coordinated vocal "displays," sometimes supported by beating great tree roots as drums. This context regularly involves the sexual access to or the captivation of female animals. If wars were often fought over access to and control over women, and if they, moreover, were (and remain) connected to an affective synchronization of the warriors by means of music, then Darwin's juxtaposition of "love" and "battle" in relation to music is more than a scholarly reference to a topical allegory brought to literary perfection by Shakespeare. Perhaps "ardor for war" knows no stronger evolutionary source than sexual competition. The same holds true for the feeling of triumph when the competition is won.

Darwin thus maps his antithetical range of music-associated emotions onto an evolutionary scenario of sexual courtship. At the same time, he stipulates a constitutional vagueness, or indeterminacy, of this reference to a sexual scenario. The feelings that are associatively activated by music often lack cognitive clarity and distinctness:

> As Herbert Spencer remarks, "music arouses dormant sentiments of which we had not conceived the possibility, and do not know the meaning"; or, as Richter says, "tells us of things we have not seen and shall not see." . . . The sensations and ideas thus excited in us by music, or expressed by the cadences of oratory, appear from their vagueness, yet depth, like mental reversions to the emotions and thoughts of a long-past age.[180]

This passage proves Darwin's capacity to take up not only much-discussed desiderata but also particularly memorable phrases in a way that opens up surprising new perspectives. Spencer's *The Origin and Function of Music*

179 Cf. Hagen, "Music and Dance as a Coalition Signaling System," 21–51.

180 Darwin, *Descent of Man*, II 336. Darwin's book *The Expression of the Emotions in Man and Animals* entails a converging passage (217) highlighting the great importance Darwin attributed to the hypothesis of latent sexual associations elicited by music.

(1858), which Darwin quotes, considers the intensification and elaboration of verbal emotional expressions as the very origin of music: "Music originates in the modulations of the voice under the influence of music. . . . It produces an idealized language of emotion."[181] For Spencer, the function of music can be directly derived from this hypothesis of the origin: music serves the cultural refinement of our capacity to express emotions of ever greater complexity, thereby supporting a development of culture away from "barbarism." Spencer's text closes with an idealist vision of a happy and even blissful cultural state in which music fulfills its mission:

> If it is the function of music to facilitate the development of this emotional language, we may regard music as an aid to the achievement of that higher happiness which it indistinctly shadows forth. Those vague feelings of unexperienced felicity which music arouses—those indefinite impressions of an unknown ideal, may be considered as a prophecy, to the fulfilment of which music is itself partly instrumental.[182]

Contrary to Spencer, Darwin's comparative chapter on "Voice and Musical Powers" argues that, while music does involve elaborate "modulations of the voice under emotion" in countless species, it still developed quite independently of language. Thus, he finds Spencer's theory of music as gradually arising from language rather improbable from an evolutionary perspective. Moreover, Darwin doubts that the mere expression of emotions can on its own turn into music:

> Mr. Spencer comes to an exactly opposite conclusion to that at which I have arrived. He concludes that the cadences used in emotional speech afford the foundation from which music has been developed; whilst I conclude that musical notes and rhythm were first acquired by the male or female progenitors of mankind for the purpose of charming the opposite sex.[183]

Spencer conceives of music in terms of expression alone, and thus in terms of the sender. Darwin regards this as insufficient to explain the duration and elaboration of musical performances. He stipulates another powerful process of what is today called "signal selection," one that proceeds from the perceiver. Clearly, a mere expression of a peacock's feelings would have never produced his fan. Analogously, in Darwin's view, which is

181 Spencer, "The Origin and Function of Music," 376.

182 Ibid., 383.

183 Darwin, *Descent of Man*, II 336, footnote.

widely accepted today, courtship-associated music in the animal kingdom evolved thanks to an increasingly extreme signal selection on the part of the courted sex—and not primarily because the "courtier" sought to fine-tune the expression of his own feelings ever further. This means that both bodily ornaments and artful displays result from an ongoing interactive evolution of the trait in question and of the preferences of the evaluators. These dynamics are not categorically different from the interaction between the artists' efforts and their appreciation by onlookers/customers, which drives the dynamics of human art history. The difference is only gradual in that the animal arts involve ontogenetic learning (change) along with powerful processes of genetic change, whereas the dynamics of the human arts can to a greater degree be explained by ontogenetic (cultural, historical, social) variance only, even though they also have genetic components.[184]

Darwin does not challenge Spencer's insights into similarities between emotional voice modulation, emotional language, and emotional music. Rather, he complements these with a contiguity association that Spencer could all the easier disregard as he did not consider animal singing: namely, the bond between "musical powers" and sexual courtship. This feedback leads to a reversal on the temporal axis of the evolution of music and language: music predates language rather than following it; and instead of prophesying and bringing about an ideal cultural future, it is a power that preserves, and connects us back to, ancient sexually charged resources of powerful emotions.

At precisely this point, Darwin notes an interesting point of convergence of his and Spencer's otherwise divergent lines of argument. Whether music is associated with an ideal cultural future or with the evolutionary past, the perception of music implies dimensions that cannot be exhaustively captured in precise cognitive terms. Darwin's intervention is worth a second round of comments:

> As Herbert Spencer remarks, "music arouses dormant sentiments of which we had not conceived the possibility, and do not know the meaning; or, as Richter says, tells us of things we have not seen and shall not see." . . . The sensations and ideas thus excited in us by music, or expressed by the cadences of oratory, appear from their vagueness, yet depth, like mental reversions to the emotions and thoughts of a long-past age.[185]

---

184 Cf. Prum, "Coevolutionary Aesthetics in Human and Biotic Artworlds," 826–27.
185 Darwin, *Descent of Man*, II 336.

"Richter" is here the German Romantic writer Jean Paul Friedrich Richter, usually known as Jean Paul. Like Spencer, Darwin read his work and cited it approvingly.[186] Still, he only uses Spencer's and Jean Paul's sentences as a descriptively convergent and highly evocative facade in order to displace the configuration of emotional depth and vagueness/elusiveness from a Romantic utopia onto a proto-myth of our forgotten sexual prehistory. At the same time, Darwin gives Kant's hypothesis of the principal indeterminacy of aesthetically awakened "ideas"[187] a novel turn. The much-discussed mystery of music's strong grip on our emotions, combined with its conceptually indeterminate character,[188] thus finds a daring evolutionary reinterpretation, one that understandably includes substantial qualifications and caveats ("to a certain extent," "if we may assume"):

> All these facts with respect to music . . . become intelligible to a certain extent, if we may assume that musical tones and rhythm were used by our half-human ancestors, during the season of courtship, when animals of all kinds are excited not only by love, but by the strong passions of jealousy, rivalry, and triumph. From the deeply-laid principle of inherited associations, musical tones in this case would be likely to call up vaguely and indefinitely the strong emotions of a long-past age.[189]

Informed by Freud, Adolf Portmann, and Konrad Lorenz, Arnold Gehlen proposed a genealogy of the visual arts that is widely compatible with Darwin's theory of music.[190] Gehlen, too, diagnoses an increasing dissociation of aesthetically potent visual "triggers" (*Auslöser*) from sexual attraction, along with the tendency that the remaining "residues of instinct" become emotionally and functionally "fuzzy." This assumption extends Spencer's and Darwin's hypotheses on the vagueness of music-induced feelings to the visual arts. According to Gehlen, it is precisely this functional fuzziness that renders aesthetic stimuli usable for a great variety of different goals, including symbolic ones. We still receive "emotional impulses" (*Gefühlsstöße*) from aesthetic stimuli, but the diffuse enjoyment of things "of an appealing gestalt" all over "the field of perception" replaces unambiguously sexual feelings.

---

186 Darwin, *Descent of Man*, I 45.
187 Kant, *Critique of the Power of Judgment*, 314.
188 Cf. Scruton, *The Aesthetics of Music*, 159–61, 360–64, 463–66.
189 Darwin, *Descent of Man*, II 336–37.
190 Gehlen, "Über instinktives Ansprechen auf Empfindungen," particularly 112–25.

For Gehlen, this "de-differentiation" of aesthetic sensitivity is associated with an "inversion of its motivational direction [*Umkehr der Antriebsrichtung*]." In the context of sexual triggers, the appearance-induced "emotional impulses" lead to a "useful change in the outside world" (i.e., mating and sexual reproduction). The "fuzzy" and "convertible" aesthetic stimuli of the human arts, on the other hand, exclusively produce a "change of one's internal state." Their crucial function consists in changing moods, emotions, and other internal states of the onlookers. In this, the experience of art resembles drug use, hypnotic dances, and social and/or religious rites, many of which also directly involve aesthetically elaborate practices.[191]

Darwin never sought support for his hypothesis of proto-music evolving through sexual signal selection by pointing to the supernormal sexual success of outstanding musicians. His view of human music does *not* restrict it primarily, let alone exclusively, to mechanisms of sexual courtship.[192] The hypothesis exclusively states that some *emotional effects* of music might be explained by reference to the assumption that associative mechanisms which originally evolved in the context of a hypothetical sexual proto-music stay in place in current music perception, if only in the form of vague and distant associations. This view leaves the question of compatibility with other functions of cultural music—be those cognitive, social, physiological or therapeutic—entirely open. Some typical objections against Darwin's theory of music[193] can be directed, at most, at his construct of our "half-human

---

191 For Gehlen, such self-directed changes are biologically meaningless and devoid of function and consequence. However, his reference to the Kantian theme of nonconceptual, disinterested, and purposeless aesthetic pleasure (German *begriffslos, interesselos, zwecklos*) is misleading. Kant never says that "heautonomous" aesthetic pleasure must be devoid of significant cognitive and affective functions for the life of the perceiver/experiencer. Gehlen, by contrast, fails to acknowledge that inward-oriented cultural techniques can have functions of their own.

192 For Darwin's contemporary readers—unlike for the present reception, which has often no basis in direct reading—it was self-evident that Darwin seeks precisely to separate human music from the sexual functions of the bird model. Cf. Stumpf, "Musikpsychologie in England," particularly 304 sqq., 316. The same holds true for Peter Kivy's short article "Charles Darwin on Music" (see 44). It is only Neo-Darwinism that often reduces Darwin's theory of human music to his standard model of sexual selection. A recent study by Kirschner and Tomasello, on the other hand, correctly describes Darwin's position in regard to "singing for sex," pointing out that this function persists in human culture only an "evolutionary vestige" (Kirschner and Tomasello, "Joint Music Making," 354).

193 Cf. Hagen, "Music and Dance as a Coalition Signaling System," 23 sqq.

ancestors" and their sexual proto-music, but *not* at his reflections on music as practiced by Homo sapiens sapiens. These counterarguments are:

1. Darwin's theory cannot explain why heterosexual listeners like music produced by singers of their own sex.
2. It cannot explain why music is performed for whole groups in exchange for payment or presents.
3. It cannot explain why humans practice and experience music much more frequently in groups than other singing animals do.

Darwin's hypothesis stipulates no more and no less than that in all these cases diffuse affective memories of sexual desire can well be part of the processing. Moreover, it is easy to imagine (1) that a love song performed by a singer of one's own sex can call to mind one's own romantic feelings by means of empathy or identification, (2) that the sponsor of a musical group performance can gain social and ultimately sexual favors from his generosity, and (3) that the distinctive performance of individuals can still be recognized in contexts of communal singing.[194]

Darwin saw no contradiction between his far-ranging assumptions regarding mechanisms of music processing shared across many species and his emphasis on the great variety of music in animals and particularly in human cultures.[195] But how could all the vastly different musical styles and individual pieces of music possibly activate residual memories of sexual courtship according to the principle of *deep association* proposed by Darwin?

Here are two suggestions. First: across all cultural differences, listening to music engages some basic mechanisms that encode the affects in question. Secondly: in singing creatures who are highly flexible in cognition and behavior as well as in phonology and prosody, the affective script of sexual courtship is not bound to specific fixed musical patterns. Instead, the deep association arises ontogenetically, as the listening to and practicing of music is often particularly intense in sexual maturation. A partly conscious, partly unconscious memory of this contiguity association between special musical practices and the emerging of sexual desire might also inform the subsequent musical experience of both individuals and groups. In fact, ontogenetic patterns of how contemporary individuals listen to music offer some support to Darwin's hypothesis. The most active and seemingly most

194 Cf. below, p. 70, 77–90.

195 Darwin, *Descent of Man*, II 333.

formative music consumption does take place in puberty and early youth (i.e., in times of sexual maturation). Moreover, music heard during these crucial years remains particularly memorable and often preserves a strong emotional value throughout later life periods. (Radio channels specializing in hits from previous decades rely on this phenomenon.)

3.3 The Heritage of Sexual Proto-Music in Language, Rhetoric, and Literature

The present dominant understanding of the evolution of human language is based on Darwin's principle of *natural* selection. Verbal language improves both cognitive and communicative capabilities (including manipulative ones). It also probably contributed to increasing the maximal group size and the cooperation level of hominids. Individuals with above average verbal capacities possibly had better chances of achieving and preserving high ranks in social groups. According to this assumption, sexual advantages are a possible consequence or implication of language-based cognitive and cooperative advantages, but not the primary motor that led to the development of language in the first place.

Darwin's hypothesis on the relationship between language and *sexual* selection has a very specific starting point and a clearly limited range. He assumes that the evolution of articulate verbal language is tied to a (concomitant or preceding) cognitive evolution that is apparently *not* found in many articulately singing species:

> It is not the mere power of articulation that distinguishes man from other animals, for as everyone knows, parrots can talk; but it is his large power of connecting definite sounds with definite ideas . . . The mental powers in some early progenitor of man must have been more highly developed than in any existing ape, before even the most imperfect form of speech could have come into use; but we may confidently believe that the continued use and advancement of this power would have reacted on the mind by enabling and encouraging it to carry on long trains of thought. A long and complex train of thought can no more be carried on without the aid of words, whether spoken or silent, than a long calculation without the use of figures or algebra.[196]

196 Darwin, *Descent of Man*, I 54, 57.

Thus, a suitable vocal tract, high neural control of articulatory organs and the capacity for vocal learning do not in themselves presuppose the development of articulate language. Like music, language requires not only all these characteristics, but also additional *cognitive capacities*. This diagnosis has a clear negative consequence: the cognitive achievements of language *cannot* be fully ascribed to the hypothetical musical-emotional prehistory of language as presented in Darwin's model of sexual selection.[197] As a result, human language cannot be regarded as a mere further development of prelinguistic vocal abilities.[198] Moreover, such a hypothesis would have a hard time explaining why only one of the singing species went on to develop syntactic language.

At the same time, Darwin's multifactor model of the evolution of language—which includes body language elements such as "signs and gestures"[199]—opens up a positive possibility: the capacities and mechanisms that language shares with music could find an *expression in language* in a music-related form (particularly if musical capacities predate language). This is, in fact, the only claim that can be derived from a close reading of Darwin's hypothesis: the hypothetical proto-music left behind a footprint, a legacy *in* the evolution of language, but language did not evolve *from* music. Specifically, Darwin refers to the prosodic means of eliciting affective responses by means of recurrent sound patterns and other material properties of language:

> The progenitors of man, either the males or the females, or both sexes, before they had acquired the power of expressing their mutual love in articulate language, endeavoured to charm each other with musical notes and rhythm.[200]
>
> Conversely, when vivid emotions are felt and expressed by the orator or even in common speech, musical cadences and rhythm are instinctively used.[201]

---

197 Darwin, *Descent of Man*, I 55 f.

198 Darwin, *Descent of Man*, I 56. Stephen Mithen argues that Darwin's hypothesis on language cannot be correct as not all characteristics of language can be explained as derivatives of music (Mithen, *The Singing Neanderthals*, 26). This argument fits in neither with the general character of evolutionary processes (which are typically not just "derivations") nor with Darwin's actual claims, which are far more modest.

199 Cf. Fitch, *The Evolution of Language*, 470–74.

200 Darwin, *Descent of Man*, II 337.

201 Darwin, *Descent of Man*, II 336.

Prosodic characteristics of language, specifically of poetic language, have been regarded as analogous to though by no means identical[202] with those of music for a long time before Darwin. The enormous richness of prosodic variations supported by the use of voice, rhythm, and melody is central to *any* definition of music. In descriptions of language, though, these features have been historically less prominent. Theories of poetic and rhetorical language use—and hence of the elaboration of diction (*elocutio*) for aesthetic and affective purposes—are an important exception to this rule. Referring to this tradition, Darwin suggests that the poetic and rhetorical elaboration of verbal language draws on both the phonological features and the affective powers of a more musical prosody.

This hypothesis shows marked affinities to theories of poetry as a musical language of affect as developed by Herder and several Romantics. Herder, too, sees the shared ground of human and nonhuman language in the expression of emotions. Regarding their differences, he stipulates that, on the one hand, human language increasingly distanced itself from an animal-like vocal expression and has its own, primarily cognitive roots. On the other hand, he proposes that strong traces of the prelinguistic communication of emotions remain inscribed within the highly developed cognitive language. For Herder, it is this oblique continuation of affective registers that renders sentences *lively*. Thus, the "natural tones" (*Naturtöne*) of affect are supposed to be "not the roots as such, but the juices that animate the roots of [human] language."[203] Darwin uses the poetic and rhetorical topos of liveliness, or vividness, in a similar fashion when describing the subcutaneous presence of affective proto-music in human language: "when vivid emotions are felt and expressed by the orator or even in common speech, musical cadences and rhythm are instinctively used."[204]

In line with Herder's and Darwin's assumptions, studies in developmental psychology have shown music-analogous characteristics both of preverbal utterances of infants and of infant-directed speech. Neuroscientific research has identified parallels in the neural processing of music and language.[205] Comparative studies on prosody, rhythm, and meter

---

202 Cf. Brown, "The 'Musilanguage' Model of Music Evolution."

203 Herder, *Abhandlung über den Ursprung der Sprache*, 701.

204 Darwin, *Descent of Man*, II 336.

205 Cf. Patel, "Language, Music, Syntax and the Brain"; Patel et al., "Processing Syntactic Relations in Language and Music"; Maess et al., "Musical Syntax Is Processed in Broca's Area"; Koelsch et al., "Bach Speaks: A Cortical 'Language-Network' Serves the Processing of Music"; Levitin et al., "Musical Structure Is Processed in 'Language' Areas of the Brain"; and Koelsch, "Towards a Neural Basis of Music Perception," 578–84.

in language and music promise further interesting findings.[206] Drawing on forty poems, an experimental study on features of parallelistic prosody showed that the poetic treatment of prosody does, in fact, enhance aesthetic appreciation and drive both feelings of being emotionally moved and perceived overall intensity to higher levels.[207]

To the extent that Darwin conceptualizes language as an affective rhetoric that works with musical means, he considers it an integral part of his archeology of cultural music:

> The impassioned orator, bard, or musician, when with his varied tones and cadences he excites the strongest emotions in his hearers, little suspects that he uses the same means by which his half-human ancestors long ago aroused each other's ardent passions, during their courtship and rivalry.[208]

This sentence, the concluding summary of the whole chapter on human "voice and musical powers," gives a new crypto-sexual meaning to the *ars oratoria*, or rhetoric. Sexually seductive speech represented a classical pattern of rhetorical persuasion as early as the very beginnings of ancient Greek rhetoric. Gorgias defends Helen against the criticism of insufficient resistance with the daring argument that it is impossible to resist a good seductive speech.[209] If so, it is not Helen who is responsible, but the drug-like power of rhetorical language that uses all rhythms and sounds at its command to produce a "magical" effect. For Gorgias, this erotic and psychopharmacologic primal scene of sexually seductive rhetoric demonstrates why and by what means language is a "great ruler" (*dynástes mégas*).

Darwin replaces the primal scene of Helen's seduction by the power of speech through an imaginary, objectless memory of sexual courtship as inscribed in the poetic and rhetorical use of language. Thus, he displaces the sexual power of rhetoric from a patent scene of its use into a now-latent quality of its formal means. In his view, poetic language use, too, continues to draw affective energy from an unconscious associative reference to a long-past practice of proto-musical sexual courtship. No matter what the occasion is, when the orator uses the musical heritage of language as a material channel of his authentic or pretended pathos, he is not using objective arguments to persuade listeners but employs deeply rooted

206 Cf. Patel, *Music, Language, and the Brain*.
207 Menninghaus et al., "The Emotional and Aesthetic Powers of Parallelistic Diction."
208 Darwin, *Descent of Man*, II 337.
209 Gorgias, "Encomium of Helen," 76–84.

affective resources. Experimental studies on poetic and rhetorical sentence patterning[210] show that:

1. rhythm- and sound-based musical virtues of words are inherently pleasurable and tend to intensify emotional effects;
2. such virtues also prime the attribution of complex perceptual gestalt properties (such as beauty and melodiousness);
3. rhetorically elaborate sentences are often more difficult to understand than analogues devoid of such elaboration, and are nevertheless preferred for purposes of persuasion;
4. rhetorically artful sentences are more often falsely considered familiar in memory tests than their non-rhetorical analogues, thus profiting from the psychological effects of familiarity (simpler, positively colored processing, higher acceptance).

Taken together, all this demonstrates that the non-semantic, proto-musical elaboration of the material and perceptive qualities of language is, indeed, a highly effective "engine of persuasion," as Kant critically put it.[211]

Darwin's hypothesis also allows for another interpretation: listening to the affective music of an orator, one feels not only courted but also placed into the powerful role of the aesthetically judging and selecting party. Barack Obama's election speeches, which were largely considered rhetorically convincing and appealing, can serve as an example here. On the one hand, these speeches drew much of their power from the speaker's performance: voice, prosody, gestures, body language. On the other hand, they also employed classical rhetorical textbook figures designed to enhance emotional effects. Such political courting of voters is a direct analogy to sexual courtship by means of the proto-music of poetic language. In both cases, the audience is an object of libidinous cathexis, and the goal of appealing to them motivates the speaker to great achievements in the art of rhetoric. The emotionalizing power of the speech is a prime index of these rhetorical achievements. The resulting excitement of the listeners can then be understood as a transfer of affect from the speaker to the audience; additionally, it can be, and even is designed to be, processed as a desire *for*

210 Menninghaus et al., "Rhetorical Features Facilitate Prosodic Processing"; Menninghaus et al., "Sounds Funny?"; Menninghaus et al., "The Emotional and Aesthetic Powers of Parallelistic Diction"; Obermeier et al., "Aesthetic Appreciation of Poetry Correlates with Ease of Processing in Event-Related Potentials."

211 Kant, *Critique of the Power of Judgment*, § 53.

the speaker and would in this capacity motivate the (desired) act of choice. Both vectors of affect—the transfer from speaker to listener, and the motivational step from listening to a speaker to desiring him or her—are not diametrically opposed in as far as both sexual and political "courtships" have the goal of instilling desire for the courting party in the courted one.

Commercial advertising, too, heavily draws on such mechanisms of both transferring affect and instilling desire. In German, *Werbung* means both "courtship" and "advertisement." Commercial "courtship"—and especially multimodal advertising that blends the affective powers of images with those of speech and music in its efforts of persuasion—systematically infuses the inherent "music" of rhetoric with sexual associations. Thus, it recruits the associative pathways of (latent) desire stipulated by Darwin. For Plato, such distant surfing on the waves of sexual courtship might well be the ultimate argument for his criticism of the rhetorical "guiding of souls through words" (*psychagogia dia logon*).[212] Darwin refrains from morally judging the phenomenon in one way or another. His great merit is to provide a speculative but nonmetaphorical and rationally coherent genealogy for our strong emotional affectability by language, and, specifically, to attribute this emotional power not only to words per se, but also and even primarily to ways of elaborating the emotional prosody which partly resonates with preverbal layers of emotive vocalizations (and their hypothetical elaborations in proto-music).

Beyond observing formal similarities between music and language, Darwin assigns these similarities a functional role in supporting emotional transfers from music to language. The musical qualities of languages can be described in rhetorical categories: as rhythms, meters, sound figures of all kinds, devices of repetition and antithesis in syntax and semantics, prosody, and stylistics. Moreover, they share the rhetorical goals of eliciting esteem/admiration for the speaker and acceptance for his agenda (*conciliare*), pleasing aesthetically (*delectare*), moving emotionally (*movere*), and occasionally also of persuading (*persuadere*)—according to both traditional rhetorical theory and recent empirical studies.[213] Distinctive competence in rhetorical language use might well be (and have been) socially advantageous, be it

212 Plato, *Phaidros* 261a.

213 Cf. Mothersbaugh, "Combinatory and Separative Effects of Rhetorical Figures"; Giora, "Weapons of Mass Distraction: Optimal Innovation and Pleasure Ratings"; Peer, "The Measurement of Metre"; Menninghaus et al., "Rhetorical Features Facilitate Prosodic Processing"; Menninghaus et al., "The Emotional and Aesthetic Powers of Parallelistic Diction."

for sexual or other goals.[214] Like narrative capacities (which are far better researched), rhetorical abilities are a special part of our general language use; they, too, are likely to be subject to special ways and time-windows of ontogenetic acquisition and may even have a distinct genetic base.

In Darwin's analysis, affectively loaded musical tones, rhythms and cadences can be found throughout the various domains of language, including everyday speech. The triad of "orator, bard, or musician"[215] only stands for the highest and most marked degrees of the proto-musical elaboration of language. This theory of rhetorical-musical language use easily overcomes the difficulties the costly signal theory has with human language.[216] According to Amotz Zahavi, words lack the reliability and trustworthiness that accrues to costly signals by virtue of being costly to produce; after all, words are freely available and typically easy to pronounce, at least for native speakers of a given language.[217] Whereas the peacock cannot fake his fan, a sexually courting person can use words to promise whatever s/he wants without any effort on her/his part, and also without intending to keep the promise. Therefore, in Zahavi's view, human language use does not qualify, as metabolically costly body ornaments and difficult performances of singing and dancing do, as a "truthful" and "honest" indicator of superb fitness.

However, the veracity of a verbal declaration of love can in fact be judged—and typically is judged—in a very discriminating fashion. This judgment refers not only to the declarative content of the message, but likewise takes into account nuances of tone, pitch, voice modulation, accompanying gestures, and compatibility with both the situation and the particular person. As speech acts are extremely complex and multimodal, there are good chances of discovering signs of missing honesty in one of these many registers. In general, multimodality tends to increase the comprehensibility of a verbal message insofar as cross-modal redundancies (each modality saying more or less the same by different means) dispel possible fuzziness and ambiguities. However, when a message is critically scrutinized for ambiguities, the partial non-redundancy of the single channels can be checked for indications that justify doubts about the contents of the verbal message. Such efforts can rely on the fact that the individual channels of a verbal message differ in terms of how easily they can be intentionally

---

214 Cf. Dutton, *The Art Instinct*, 147 sqq.

215 Darwin, *Descent of Man*, II 337.

216 Cf. Steinig, *Als die Wörter tanzen lernten*, particularly 51.

217 Zahavi, *The Handicap Principle*, 80 and 221–23. See also Knight, "Ritual Speech Coevolution"; Power, "Old Wives' Tale: The Gossip Hypothesis and the Reliability of Cheap Signals."

manipulated. Truthful words can easily be substituted by untruthful ones; however, speech prosody, gestures, and body language are more difficult to manipulate in a consciously deceptive fashion. Attempting to spontaneously integrate all individual channels of an utterance into a consistently deceptive message is far from simple and can easily lead to discrepancies which, even if minimal, reveal a dishonest intention.

Zahavi's hypothesis that human verbal language is "cheap" rather than costly only makes sense if the multimodal nature of language is disregarded. However, this disregard is not even appropriate where language lacks an emotional prosody based on articulate sounds as well as an articulating body (i.e., in written literature). Literature, including prose, does not only create imaginary equivalents of an author's "voice," but also it draws much of its power from its rhythm and thus from a compositional quality that has a sensual "gestalt" property even in the absence of spoken sound. This non-semantic signature of a prose style is far from easy to imitate, let alone fake. Readers spontaneously perceive the imaginary "voice" of an author and the rhythm of his or her sentences even if they cannot analytically describe them in any adequate fashion. Poetic language is moreover often subject to complex metrical and/or other genre requirements that take great artistic powers to live up to. In this and many other respects, poetic language—and the language of a good orator—is expected to be rhetorically elaborate and to engage our attention, emotional responses, and memory in distinct ways, and not least to activate our reward system. A language that meets all these demands can hardly be a "cheap signal" freely usable for deceptive purposes at no cost.

4. Peacocks/Songbirds and Human Artists: Merits and Limits of the Parallel

Darwin's bird model of artful singing replicates all factors of the physical ornament ("beauty") model. It predicts that sustained sexual selection from the part of the courted sex can drive artistic performances to ever higher degrees by virtue of making sexual success dependent on the quality of these performances. The human arts involve an analogous feedback between art production and reception, be it through the role of patrons, customers, and market mechanisms, and through the processes of canon formation. However, Darwin nowhere predicts an effect of human art production on reproductive success. Rather, he exclusively diagnoses distant and vague traces of sexual associations in the emotional effects of singing, rhetoric, and poetry, and interprets these as an indication that earlier in the evolution of humans also the functional implications of the animal model may well have been in place. As a result of this limitation of the parallel, it did not occur to Darwin to seek statistical evidence that artistic success translates into reproductive success.

Many mythologies of artists conform to the predictions of Darwin's model by combining two seemingly antithetical features: they portray artists as sexual attractors or directly parallel artistic and sexual acts, but they do not include any reference to high numbers of children. Orpheus is perhaps the most renowned example.[218] His musical performance clearly makes him a supernormal sexual attractor. The sound of his cithara has the effect of a tender caress (*mulcens*) upon Eurydice.[219] And when Orpheus finally loses Eurydice, his sad music inspires intense sexual desire in many women,[220] so much so that one can legitimately read this part of the myth as the first portrayal of the groupie phenomenon. Women get a crush on the singer collectively and simultaneously; aesthetic and sexual attraction converges.[221] In hyperbolic intensification of its power to win over and emotionally affect its listeners, Orpheus's music can move even gods, birds, and fish. The myth stresses the sexually persuasive powers of music all the more as Orpheus's own intentions after losing Eurydice are anything but sexual. He merely wishes to commemorate his loved one by singing. But even this deeply sad song provokes ardent sexual desire in his female listeners. Their passion is so tempestuous that they cannot bear to be rejected by the singer, who is subjectively uninterested, but appears to "objectively" court them qua music. Raging in ersatz intimacy, they tear him to pieces.

Thanks to his early death, Orpheus is transfigured into a mythological ideal of both a perfect singer and a consummate lover. However, sexual desirability and reproductive success—features that should correlate positively according to the evolutionary model—are radically dissociated in the case of this proto-singer: Orpheus dies without offspring. Despite this important limitation, the Orpheus myth—like the modern cult of pop stars—shows a good fit with Darwin's model of male aesthetically exuberant courtship and female choice. In current fan culture, too, female fans are more inclined to respond with fairly direct sexually colored excitement, indeed arousal, to the performances of male stars than are male audiences

218 Cf. Ovid, *Metamorphoses* XI, 1–66.

219 Hyginus, Fabulae 164.

220 Cf. Ovid, *Metamorphoses* X, 81 sqq.

221 The sirens' song exerts analogous effects with inverse gender roles; however, it is less clearly identified as a sexual attractor.

in response to female singers.[222] If this phenomenon could be shown to be transculturally homologous and not explainable by upbringing and other cultural influences, it could be seen as a trace of ancient practices of male singing and female choice. (However, conclusive proof of such a relationship seems hardly possible.)

Goethe provides a variation on this theme in his poem about the "Rat-catcher of Hamelin":

> Sometimes the skillful bard ye view
> In the form of maiden-catcher too;
> For he no city enters e'er,
> Without effecting wonders there.
>
> However coy may be each maid,
> However the women seem afraid,
> Yet all will love-sick be ere long
> To sound of magic flute and song.[223]

The singer is a "maiden-catcher"—the singing for sex-hypothesis could barely be stated more bluntly. At the same time, Goethe's poem leaves it open whether the lovesickness elicited by the singer in his female audience is more of a phantasmatic nature or actually leads to direct sexual encounters with him, let alone to a long line of descendants fathered by the singer. The poem exclusively focuses on the very moment of being affected by the "sound of flute and song" and says nothing about real consequences deriving from this affection.

For Plato, too, art production and eros are closely associated;[224] yet, again, this does not imply that good artists leave behind more children than less talented artists and nonartists, but only that their devotion to a different kind of progeny is as passionate as that of a sexual lover. Pygmalion is another mythological example for the strained relations between artistic, sexual, and reproductive success. Ovid writes that Pygmalion had no interest for the real women of his times.[225] Instead, he creates an ideal woman of stone. He talks to her, gives her presents, and subsequently touches, kisses, and caresses the statue, thus treating it as a genuine sexual object. Finally,

---

222 Cf. Fritzsche, "Fans und Gender," 238–41; Fritzsche, "Pop-Fans. Studie einer Mädchenkultur."

223 Goethe, "The Rat-Catcher," 108.

224 Plato, *Phaidros*.

225 Cf. Ovid, *Metamorphoses* X, 243–97.

he asks Aphrodite to bring the stone to life; she agrees, and he performs a sexual act with the statue the very moment it becomes alive. Many ancient writers ridiculed this behavior.[226] The more idealistic and metaphorical reading—Pygmalion as the epitome of the capacity to be strongly affected by aesthetic illusions—tends to forget that Pygmalion is clearly portrayed as a precarious creature: disturbed relationships with real women, narcissistically and incestuously colored sex with a self-created female image, a fatal tendency for (further) incest in his offspring (Myrrha), and, finally, the extinction of the whole line with the death of Adonis, who cannot be seduced even by Aphrodite herself.[227] Doubtlessly, Pygmalion celebrates a great triumph as an artist when stone is turned into flesh. His achievement as a sculptor is therefore regarded as a pivotal example of aesthetically "lively" and "enlivening" creation.[228] Still, Pygmalion, too, is certainly not an example for above average reproductive success in the sense of evolutionary theory; this is all the more noteworthy as his narrative almost identifies artistic success with a sexual act.[229]

Accordingly, one of the most powerful cultural imaginations of art—the idea of posthumous fame—conceptualizes artistic practice not as a means of self-advertising and thereby enhancing sexual success, but as an equivalent, substitution, or even the preferred alternative of an afterlife achieved by sexual reproduction. After all, posthumous fame dependent on artistic achievement can well be more durable for an individual than leaving behind a long line of descendants. Plato calls both forms of self-reproduction—artistic and sexual—*éros* and *tókos en kalô*, desire for and reproduction in a beautiful body/work of art (literally, *birthing in the beautiful*).[230] The double meaning stresses the analogous relationships of progenitor/author and his/her product in biological and cultural procreation. Both processes ensure a posthumous afterlife. Plato famously remarked that works of art are preferable to children in this regard, as nobody has ever gained a monument, much less eternal fame, for leaving behind a child, but only for great artistic achievements.[231]

---

226 Arnobras, *Adversus Nationes* VI, 22; and Clemens von Alexandria, *Protrepticus* 57, 3.

227 For a more extensive treatment of Greek and Roman myths of beauty altogether, see Menninghaus, *Das Versprechen der Schönheit*, 13–65, particularly 33–35 and 40–57.

228 Cf. Mülder-Bach, *Im Zeichen Pygmalions*.

229 Cf. Hagen and Bryant, "Music and Dance as a Coalition Signaling System," 23, and Catchpole, *Birdsong: Biological Themes and Variations*.

230 Plato, *Symposium* 201d-202b; Ritter, *Fragmente aus dem Nachlaß eines jungen Physikers*, Fr. 495–97, 504.

231 Plato, *Symposium* 209b. Cf. Horaz, *Carmina* 3, 30.

Richard Dawkins gratefully integrated this argument into his comparative typology of biological evolution qua genes and cultural evolution qua memes, and he reinforced it by pointing to the permanent remixing of genes in sexual procreation, which leaves an author's genes far less stable across generations than a well-preserved artwork.[232] Romantic concepts of artistic *generatio* have even programmatically defined the creation of art as the male response to the female privilege of giving birth: "The woman gives birth to humans, the man to art."[233] In this view, too, the number of artistic offspring would need to be counted in terms of artworks, but not children.

Summing up, the cultural record is rich in imaginations that conform to Darwin's take on the human arts in a twofold fashion: they evoke associations of artistic success and sexual attraction, but they are far from confirming the notion that artistic success literally translates into higher numbers of physical offspring. This might be read as indicating, in line with Darwin's hypothesis of an evolutionary vestige, a more phantasmatic nature of the sexual association.

Reading Darwin's detailed and nuanced reflection on the human arts against the grain, Geoffrey Miller claimed that a straightforward singing-for-sex hypothesis along the lines of Darwin's animal model also applies to human artists, citing some modern artists and pop stars as evidence.[234] In fact, Modigliani, Picasso, Mick Jagger, and Jimi Hendrix are famous for their above-average success with the opposite sex. But how well-founded and representative is this phenomenon, which is hardly imaginable without its amplification through modern mass media?

To start with, Darwin's model predicts that artistic capacities—if they can be explained as driven by sexual selection—should emerge during sexual maturation and be particularly highly expressed throughout the subsequent adult phase of intense sexual courtship. The bird arts of singing, dancing, and advantageous self-display clearly meet this expectation. Human teenagers of both sexes, too, readily learn to dance skillfully and gracefully (even without professional training) and also to sing beautifully (apart from possible constraints and disadvantages in males when the voice is breaking).[235] These capacities often appear to increase further in those years after puberty when sexual activity reaches peak levels. However, contrary to the standard animal model, these capacities, particularly those

232 Dawkins, *The Selfish Gene*, 189–201.

233 Ritter, *Fragmente aus dem Nachlaß eines jungen Physikers*, Fr. 495.

234 Cf. Miller, *The Mating Mind*, 73 sqq.; Miller, "Evolution of Human Music through Sexual Selection," 331. For a more detailed discussion of Miller's hypothesis see Davies, *The Artful Species*, 124–26.

235 Cf. Miller, "Evolution of Human Music through Sexual Selection," 354.

for singing, can well be pushed to higher levels even after the period of greatest sexual activity. Still, the overall picture lends some support to the idea that singing and dancing as cultural arts might be derived from an archaic adaptation for displays of sexual courtship.

Accordingly, modern casting shows tend to focus on looks, singing, and movement (the three domains of bird aesthetics), and there are countless young stars in pop music and the performing arts (particularly, movie actors and actresses). Lyrical poetry—being a songlike hybrid at the crossroads of music and literature—appears to be (or at least to have been, historically) an exception to this rule. Readable (love) poetry is often written by very young individuals, who give up all poetic ambitions as soon as they become adults. By contrast, in the visual arts and in the literary genres of narratives and drama, artistic mastery tends to be only reached at an age when the physical luster of youth is fading or has already faded. In particular, painting and sculpture—art forms that require more special technological skills and depend on material vehicles other than the voice and the body of the artist—only show very limited conformity with the age prediction derivable from the theory of sexual selection. Is it enough to point to the extended period of reproductive capacities of human males in order to explain arts that are typically perfected at a more advanced age as, say, special courtship devices of older men who are past their prime of youthful looks?

As to hard statistical facts regarding a positive correlation between artistic performance and sexual success across times and cultures, nothing coming close to it is available. A convincing large-scale statistical study on this correlation would be quite a novelty and a great challenge, particularly for the archaic times in which the behavior is supposed to have evolved. Social Anthropology does offer some reports of the sexual attractiveness of young men who are good at singing, in line with news reports and gossip regarding modern pop stars.[236] At the same time, a cursory glance at the lives of artists suggests that even among the authors, musicians, and painters better known today, the great majority led a modest life in material—and, it seems, also sexual—terms. For millennia, most sculptors and painters who adorned religious sites and the palaces of the upper classes remained anonymous in the cultural record, because they were largely regarded as craftsmen. Accordingly, the Greek and Roman tradition did not even linguistically distinguish between what today are considered the "fine arts" and the arts of carpenters, plumbers, etc.; all these professions

236 Cf. Hagen and Bryant, "Music and Dance as a Coalition Signaling System," 23, and Catchpole, *Birdsong: Biological Themes and Variations*.

were equally called *téchnai* or *artes*, with no special term available for the presumably "higher" arts. Typically, all these artists offered their special expertise to male patrons rather than displaying it for the opposite sex.

To be sure, a very small number of the sculptors and painters among these "artists" in a very broad sense achieved special social recognition, and this may also have translated into greater sexual success. Still, being an artist in this sense was barely a predictor of outstanding social and sexual success when compared to being born into the higher classes for which the "artists" almost invariably worked for very low wages. The German expression *brotlose Künste*, breadless arts, reflects this image. By and large, the picture appears to be not categorically different for artists in the more prestigious modern sense. The few celebrities and sex symbols among the many artists in precarious living conditions barely suffice to literally subsume the human arts under Darwin's animal model. At the same time, they do suffice to give the topical association of artistic and sexual success a modern continuation. In this perspective, Darwin's hypothesis that such associations may reflect an evolutionary vestige, a distant memory of a long-past age, may be less far-fetched than the claim that the human arts literally conform to Darwin's animal model of singing for sex.

On a final note, Richard Prum has proposed that Darwin's model of the ongoing coevolutionary feedback between minor variations in animal art production and the selective evaluation of these animal arts by the opposite sex is essentially no different from the analogous feedback loop between human art production and its evaluation by audiences, sponsors, or the marketplace.[237] Stated in this abstract fashion, the analogy clearly applies. However, beyond the fact that the evaluating part regarding the human arts is not at all categorically limited to the opposite sex, some other important systemic differences apply that Prum left unaddressed. In a stricter sense, an ongoing synchronization of artistic production and aesthetic preferences of audiences applies to the animal arts only, but not to the human arts. Aesthetic preferences of animals are always and invariably "up to date," exclusively favoring the most recent fashion of songs and other artful displays.[238] Humans, by contrast, can cultivate multiple preferences, including preferences for artworks that date back hundreds of years, and, by implication, admiration and enthusiasm for artists who are no longer alive. In fact, a great percentage of contemporary humans do not like the most recent art

237 Cf. Prum, "Coevolutionary Aesthetics in Human and Biotic Artworlds."

238 For a striking empirical proof of this assumption, see Derryberry, "Evolution of Bird Song Affects Signal Efficacy: An Experimental Test Using Historical and Current Signals."

production in painting, sculpture, poetry and other fields at all. To be sure, sexual maturation in humans is also a time where the most recent trends strongly prevail, particularly regarding songs and the fashionable practices of dressing and enhancing one's outer appearance. Still, the potential for being out of sync with the most recent coevolved state of art production and preferences—or for enjoying multiple past art-worlds—is enormous in quantity and incredibly rich in qualitative variety.

The animal model of the tight ongoing feedback between artistic displays and aesthetic preferences does not entail any provisions for this partial decoupling of aesthetic preferences from an exclusive focus on the most recent state of affairs. Such decoupling would be flagrantly dysfunctional in terms of sexual attraction,[239] and it would require capacities that are thus far unavailable to birds and other animals: namely, to create physically enduring artworks or artworks that are handed down to posterity through either collective memory or technical storage media. These and other differences between the animal and the human arts—including the biologically evolved human adaptations on which they rely—are the subject of Chapter 4.

239 Cf. Dutton, *The Art Instinct*, 235.

2

The Arts as Promoters of Social Cooperation and Cohesion

This chapter discusses hypotheses that propose the very opposite of Darwin's evolutionary narrative: the arts are not competitive efforts aimed at promoting the differential (reproductive or symbolic) success of individuals relative to other individuals; rather, they precisely enhance social cooperation and cohesion. To be sure, not all such hypotheses propose that this cooperative function may have driven the very evolution of the arts, but some do.

The first section discusses Darwin's distinction of calls and songs as projected onto John R. Krebs's and Richard Dawkins's typology of competitive versus cooperative animal signaling.[1] In this view, calls are cheap and cooperative signals, whereas songs are costly and competitive. This conceptual framework will be shown to place narrow constraints on the hypothesis that infant-directed speech may have been the first human art of singing.

Subsequently (Section 2), two hypotheses will be discussed that go beyond Krebs's and Dawkins's clear-cut dichotomy of cheap cooperative and costly competitive signaling in that they suggest types of human signaling that combine the two features "costly" and "cooperative." One of these hypotheses proposes cooperative benefits of costly artful displays for relations *between* different groups of humans,[2] the other for the social

1 Krebs and Dawkins, "Animal Signals," 390 92.

2 Hagen and Bryant, "Music and Dance as a Coalition Signaling System"; and Hagen and Hammerstein, "Did Neanderthals and Other Early Humans Sing?" Cf. also Geissmann, "Gibbon Songs and Human Music from an Evolutionary Perspective."

relations *within* groups. For both cases of costly signals of cooperation, it will be argued that they are, in fact, not logically incompatible with Darwin's and Krebs's/Dawkins's dichotomy, but add a *tertium datur* that may only be found in human behavior.

1. The Arts as Costly, Competitive Signals, and the "Motherese" Hypothesis

Drawing on the *costly signal* theory,[3] Krebs and Dawkins have proposed a typology of competitive and socially cooperative signaling.[4] In this view, only competitive and manipulative signals are costly for the sender. For instance, a sender has an interest in frightening off, deceiving, manipulating, or winning over other individuals. In principle, the receivers of the message do not necessarily share the interests of the sender. Therefore, in an evolutionary arms race, receivers tend to increase the costs of manipulative signals for the respective senders, such that these end up entailing some information about the sender's qualities and the "honesty" of his signals. These costs act as constraints on cheap ways of cheating. From such selection pressures exerted by the recipients of signals, there result the numerous costly signals that can be observed in animal communication. In the case of vocal signals, many potentially demanding signal characteristics (length, intensity, complexity, rhythm, melody, etc.) have been evolutionarily selected that enable the receiver to draw conclusions about the sender's capabilities.

Cooperative signals, on the other hand, tend to be inconspicuous and "cheap" because and to the extent that the cooperating partners have the same interests in the respective situations. Under this premise, the elaboration of the signal provides no competitive advantages. This is why songbirds do not respond to the appearance of a predator with a complex aria, but rely on simple, short calls with relatively low production costs. In such cases, no interest is at stake that could be promoted by high signal-production costs. Moreover, "once an alarm call is recognized as an alarm, its acoustic form and content is not subject to sensory evaluation and judgment."[5] The latter, however, applies to songs and other costly signals. Here, the form itself is

3 Cf. Zahavi, "Mate Selection: A Selection for a Handicap"; Zahavi, "Decorative Patterns and the Evolution of Art"; Zahavi and Zahavi, *The Handicap Principle: A Missing Piece of Darwin's Puzzle.*

4 Krebs and Dawkins, "Animal Signals," 390–92.

5 Prum, "Coevolutionary Aesthetics in Human and Biotic Artworlds," 816.

evaluated for artistic qualities, and consistent aesthetic preferences on the part of the receiver, just like individual differences on the part of the performer, drive the further evolution of the signal in a self-reinforcing loop of feedback between traits and preferences.

Such coevolutionary processes likewise apply to the colors of flowers which function as advertisements to pollinators who, in turn, drive the evolution of this "biotic art."[6] In this case, the selection for aesthetic appeal is exerted by other species and not of a sexual nature. In contrast, animal body ornaments, songs, and dance displays are subject to a within-species selection, and, specifically, to a selection by the opposite sex. Human artworks share with animal art their dependence on within-species evaluation; however, even though many human artworks are also of a gendered nature, they are typically not specifically evaluated and selected by the opposite sex only. In fact, much of human "high" art was produced by male artists for male patrons, indicating that their advertisement function was more about power and prestige than about a direct sexual selection of the artists. (Still, the patrons might also have sexually profited from adorning themselves with beautiful buildings and ornaments.)

Getting back to Krebs's and Dawkins's typology of cheap and costly signals, it includes a clear rule for assigning the two types of vocal signals distinguished by Darwin to two groups: calls have the robust, simple, and easily decodable acoustic characteristics of cheap cooperative signals; courtship songs, on the other hand, feature the complex acoustic characteristics of costly competitive signals. To be sure, Darwin's domain-specific typology of calls and songs does not entail the assumption that sexually selected costly signals, such as elaborate artistic performances, are indicators of heritable fitness in the sense of natural selection. At the same time, it does by no means explicitly rule out a possible substantial overlap of sexually selected traits with naturally selected fitness.[7] In any event, Krebs's and Dawkins's typology is used in the following primarily as one that descriptively converges with Darwin's distinction of calls and songs, while overcoming the limitation of Darwin's typology to a specific sensory domain. A natural selection–based association of costly signals with "good genes" is not automatically implied in this use of the distinction between two types of signals.

The typology of cheap versus costly signals can be used as an argument against the hypothesis that the apparently pan-cultural praxis of early prelinguistic mother-child communication may be the origin of human

6 Cf. ibid.

7 Cf. Darwin, *Descent of Man*, I 256.

music.[8] Heightened pitch and higher rhythmical regularity are universal features of infant-directed speech (IDS), or "motherese," and thus perhaps not learned behaviors. IDS and lullabies usually do not require extensive practice, carry relatively clear emotional messages in their prosody ("I'm here, everything's fine, and safe, no need to worry"), and do not tend to compete in virtuosity. Also, the exclusive addressee of the singing is usually not subject to competition between mothers. Thus, IDS and lullabies do not demonstrate the typical features that are dependent on a strong competitive component (i.e., high complexity, high individual [and cultural] variance, and a high dynamics of change). Therefore, the vocalizations involved in mother-child communication can far better be assigned to the category of species-typical coordination calls than to the category of songs and highly artful music.

This typological classification does not at all preclude the possibility that the special art of singing and speaking characteristic of IDS constitutes not only an ontogenetic but also a phylogenetic basis of our capacity to create and perceive music. However, it does suggest a distinction, in Darwin's sense, between the hypothetical evolution of our vocal capacities for a variety of pragmatic purposes, on the one hand—and, on the other, the *elaboration* of these evolved capacities into a competitive "art" of "*high* musical development."[9] In line with Darwin's distinction, only productions of this latter type, but not motherese and lullabies, are regularly lauded for their high aesthetic appeal.

2. Artful Multimedia Performances as Costly Practices of "Courting" Preferred Allies?

According to Krebs's and Dawkins's typology of cheap and costly signals, elaborate songs and other aesthetic practices can be expected to always serve individual competition for non-sharable resources, but not social cooperation. Supporting this notion *ex negativo*, communal singing has been interpreted as "a cheap and easy form of interaction": apparently, everyone can join in without any major "costs," even while at the same

---

8 Cf. Dissanayake, "Antecedents of the Temporal Arts in Early Mother-Infant Interaction"; Dissanayake, "The Arts after Darwin"; Falk, "Prelinguistic Evolution in Early Hominids: Whence Motherese?"; Unyk et al., "Lullabies and Simplicity"; Trehub et al., "Parents' Sung Performances for Infants"; Trehub et al., "Infants' and Adults' Perception of Scale Structure"; McDermott and Hauser, "The Origins of Music," 33 sqq.

9 Darwin, *Descent of Man*, II 335.

time engaging in some useful activity other than singing.[10] However, this view underestimates the cumulative effort of memory and learning needed to master the culturally specific rhythms, harmonies, and texts to such a degree that spontaneously joining in becomes easy. Even more strikingly, many social rites—while unquestionably supporting social cooperation/cohesion[11]—can hardly be interpreted as "cheap signals." The dances performed at such gatherings need to be learned and practiced extensively; musical instruments need to be produced and their usage needs to be learned, both of which is highly time-consuming; texts and lyrics must be memorized, etc. Thus, joint music making and dancing of this type transcend the dichotomous identification of cooperative with cheap and of competitive with costly signals.

For Hagen, Hammerstein, and Bryant, elaborate musical-theatrical performances of social groups evolved as joint displays for other groups rather than as practices of within-group bonding. The authors propose that the goal of such performances was to impress select groups that were targeted as preferred allies, and hence to pave the way to a mutual agreement of alliance. If several human groups compete for alliances with the same target group, then the goal of cooperation itself includes competition and hence favors costly rather than cheap signaling. In such cases, social cooperation is the goal and function of the behavior, just as in simple calls. However, contrary to motherese and other relatively non-costly cooperative signals, the desired cooperation can only be arrived at through a competitive display.

While there is no provision for this type of signaling in Krebs's and Dawkins's taxonomy, it is, in principle, compatible with this taxonomy. After all, it does not challenge the two basic antithetical types of signaling (cheap/cooperative and costly/competitive), but adds a rare third type. Hagen and Hammerstein note that apparently only humans build *alliances across groups*, thereby underlining the exceptional nature of such behavior. Through coordinated artful performances directed at potential allies, human groups presumably showed off their own "coalition quality," just like sexually courting birds display their "partner quality" to potential partners through their capacities for singing:

10 Mithen, *The Singing Neanderthals*, 214.

11 Freeman, "A Neurobiological Role of Music in Social Bonding." Cf. also Blacking, *How Musical Is Man?*; Benzon, *Beethoven's Anvil: Music in Mind and Culture*; and McNeill, *Keeping Together in Time.*

> Choosing allies based on the quality of music and dance performances (as well as the quality of other forms of cultural production prominently displayed at feasts, such as food, clothing, artifacts, etc.) would spark an evolutionary arms race between coalition members with an interest in producing ever more convincing signals of coalition quality, and potential allies with an interest in better discriminating between performances of coalitions of different quality, leading, eventually, to the rich coalition signaling system we call music.[12]

According to this hypothesis, the capacities for social coordination displayed in complex musical performances are themselves the message of these displays: they demonstrate exactly the capacity for joint actions that make a group eligible as a superb ally. At the same time, the hypothesis that music actively promotes social cohesion *within* the performing group is explicitly rejected.[13]

However, the model can easily be read against the grain. If performances in the presence of alliance partners often require months or even years of practice (as the authors stipulate), then the *group-internal* practicing takes up much more time and energy than the (hypothetical) performances in the framework of alliance rites. In this view, the intragroup advantages of such practices—in terms of *intra*group cooperation—might even be larger than the benefits that might accrue from their one-time performance for individuals from other groups.

It is also questionable whether or not the majority of elaborate musical and dance performances in traditional cultures were actually performed before outgroup members. The sources cited by Hagen and Bryant are not nearly sufficient for a reliable quantitative statement.[14] Moreover, building alliances and exogamous mating strategies regularly and closely correlate; in archaic cultures—and also partly in less archaic cultures—the hypothetical rites of alliance building could simultaneously serve the function of a marriage market.[15] Collective singing, dancing, and music making at such events would then not replace the evaluation of an individual's quality as a sexual partner with the evaluation of a group's quality as an alliance partner, but serve both purposes at the same time.

---

12 Hagen and Hammerstein, "Did Neanderthals and Other Early Humans Sing?," 10.

13 Hagen and Bryant, "Music and Dance as a Coalition Signaling System," 30.

14 Cf. also Kirschner and Tomasello, "Joint Music Making," 355.

15 Cf. Hagen and Bryant, "Music and Dance as a Coalition Signaling System," 40.

3. Joint Music Making and Multimedia Performances as Promoters of Intragroup Cooperation/Cohesion

Ethnological theories of the human arts tend to focus on their role for ethnic and cultural identity, and, by implication, for social cohesion. This focus entails the at least implicit assumption that elaborate artistic performances of humans may be costly signals that first and foremost promote *within*-group cooperation and cohesion. How can this assumption be reconciled with the standard model of costly signaling? Which special conditions must be fulfilled in order for costly "arts" to be instrumental for supporting social cooperation even within groups?

The answer suggested here is the following: social events and practices that are costly not only in terms of time and material resources required, but also of artful performances involved can promote social cooperation and cohesion if the membership in a certain social group, or in a particular stratum of this group, is not a simple unproblematic fact but involves *challenges* and *conflicts*. In such cases, group membership repeatedly calls for active investments, expenditures, and perhaps even sacrifices in the service of displaying adherence to the group's values. A cooperation-based social structure might be in need of such costly self-assurance particularly in cases where membership in a special (sub-)group is not only defined by shared interests, but also involves structural conflicts, such as non-convergent interests and changing alliances within groups, rises and falls in pecking orders and hierarchies, etc. This seems to be the case in human societies.

Thus, as in the case of the hypothesis of courting alliance partners—and contrary to alarm calls and IDS—there is a conflictual and competitive background involved in elaborate artful practices that hypothetically promote social cooperation and cohesion *within* groups. Real deeds are not the only currency that tests and proves pro-social expectations in the case of human cultures. There are also symbolic practices (in particular, social and religious rites of all kinds) that serve to build, affirm, and also test the readiness for group-conforming behavior. Such practices can, indeed should, be aesthetically costly; the participation would not have any signal function whatsoever if it did not "cost" the participating parties anything. Under these premises, elaborate self-ornamentation and performances involving dancing and music, as well as unusual/paranormal language use, could evolve as distinctive characteristics of social rites in which groups symbolically consolidate their own fundaments, values, and behavioral rules.[16]

16 Cf. Irons, "Religion as Hard-to-Fake Sign of Commitment"; and Sosis, "Why Aren't We All Hutterites? Costly Signaling Theory and Religious Behavior."

Singing, dancing, and joint music making imply continued processes of mutual coordination.[17] Nonhuman primates cannot be brought to move, much less to vocalize, in sync to a given beat. Humans are able do this rather easily, even as children. Apparently, they are the only primates able to flexibly adjust their movements and vocalizations to a great variety of meters and rhythms prescribed by songs and instrumental music. It is unlikely that this capacity only evolved to provide a new stage for competitive *individual* performances.[18] Rather, joint singing and dancing gives the abstract factual "We" of group membership a concrete event value; it is inherently self-rewarding and may well motivate transfers of cooperative behavior from symbolic actions to everyday social life. In the case of rites, sharing and participation tends to involve both production and reception; in other cases, it involves reception only. In any event, if considering the great cultural variance and the enormous costs of acquiring the high skills often required for elaborate musical performances, it can hardly suffice to regard such performances as a cheap and useful practice of social coordination.

Religious practices, which appear to be a transcultural human characteristic, correspond to the theory of aesthetically demanding symbolic acts as presented here. Religions that promise all kinds of good things but do not ask for anything in return do not seem to be taken seriously. Apparently, no such religion ever succeeded in the long run. Religions that are cognitively, emotionally, and economically demanding/costly, on the other hand, tend to reduce the conflict level within social groups by instilling shared values and behavioral codes and promising rewards for membership and participation. Moreover, the religious adoration of higher beings includes a scenario that is closely reminiscent of sexual courtship and thus suggests costly aesthetic elaboration. Like sexual partners, gods are courted for their favor (benevolence, mercy, help) by means of different signals that are costly in terms of time, discipline and, occasionally, finances: repeated professions of veneration and adherence, vows of fidelity, and material sacrifices.

The relationship between religious, political, or moral "messages" and their intensification through the arts can be interpreted in three ways: (1) the messages are primary, and the arts evolved only later in the service of

17 Koelsch, "From Social Contact to Social Cohesion—The 7 Cs."

18 Cf. Brown et al., "An Introduction to Evolutionary Musicology," 17; Geissmann, "Gibbon Songs and Human Music From an Evolutionary Perspective," 118; Molino, "Toward an Evolutionary Theory of Music and Language"; Richman, "How Music Fixed 'Nonsense' into Significant Formulas: On Rhythm, Repetition, and Meaning"; Merker, "Synchronous Chorussing and Human Origins"; Freedman, "A Neurobiological Role of Music in Social Bonding."

supporting the spread and acceptance of these messages; (2) aesthetic elaborations are co-emergent with the religious, political, or moral messages; (3) practices of aesthetic elaboration are older than religious messages and constitute necessary preconditions for their success.

A comparison with other species tentatively supports the third assumption. Coordinating calls and other motor-affective coordination mechanisms in birds, "warring" chimpanzees, and other species appear to have evolved in complete separation from transmitting a symbolic meaning. For human cultures, extant archeological findings suggest that aesthetic practices might have predated religious ones, since available evidence of practices of self-ornamentation is older than the first clear indications of religious practices. However, as narratives could not possibly have survived for thousands of years in the absence of writing, this archaeological evidence is far from conclusive. Under these auspices, it may, for the time being, be more meaningful to consider aesthetic and religious practices as different but frequently interacting forms of human culture that both create, or involve, symbolically significant objects, figures, spaces, acts, values, songs, and narratives, and that both allow for joint ways of engaging all these objects, songs, narratives, etc.[19]

The phenomenon of religion also exemplifies the downside of promoting social cohesion and cooperation that is supported by aesthetic appeal. Precisely because a social group establishes and reinforces shared love for its own gods at great costs, thus supporting intragroup cohesion, this very same group can turn—with all the greater motivational power and efficiency—against other social groups worshipping other gods and employing other aesthetic practices. Throughout human history, strong cooperation and cohesion within a group appears to correlate closely with the potential for and the tendency to aggression against other groups.[20]

Ellen Dissanayake's understanding of arts and religious rites does not sufficiently take into account the conflict-laden background and the violent potential of community-building religious and aesthetic practices. Rather, she largely separates ritual and artistic elaboration from all competitive mechanisms and interprets the costs of "making special" only

19 See also Dissanayake, "The Arts after Darwin," 257.

20 Cf. Miller's sarcastic observation: "Group competition replaces the logic of murder with the logic of genocide. Not a great moral improvement. Group selection models of music evolution are not just stories of warm, cuddly bonding within a group; they must also be stories of those warm, cuddly groups out-competing and exterminating other groups that don't spend so much time dancing around their campfires" ("Evolution of Human Music through Sexual Selection," 351). See also Pinker's critique of the social cohesion hypothesis ("Towards a Consilient Study of Literature," 173 sqq.).

as an indicator of caring for a shared issue considered to be important.[21] This focus on "caring" and "sharing" has a somewhat idealistic note.[22] Still, artworks have contributed and continue to contribute invaluably to the construction of a social, political, and metaphysical imaginary that allows for participation far beyond concrete practices of joint music making or attending elaborate multimedia rites.

Songs, dances, myths, and other narratives are not singular entities in space and time; rather, they are *virtual* objects, or scripts for performances, that can be enacted by different persons and in different contexts. Unlike everyday consumables, artworks are, in principle, not subject to one-time consumption but entail a promise of potentially eternal existence (or repeated performance) and indefinitely renewable pleasures of engaging in them. To be sure, there are great inequalities in the ownership of valuable artworks and close correlations between social status and aesthetic preferences in Bourdieu's sense.[23] Nevertheless, *social shareability is quite generally and in principle an elementary distinctive characteristic of many human artistic practices, even if humans seek exposure to them completely on their own and without a context of direct social interaction.*[24] Food resources allow for a cooperative partitioning and allocation to many individuals; still, it is impossible that many individuals consume the very same piece of food. In contrast, the reception of aesthetic performances and objects allows for precisely this: they can be fully "consumed" by all members of a group, without differential allocation of parts of these performances and objects. In this sense, artworks of all sensory domains constitute an *invaluable resource of indefinite social shareability.*

At the same time, artworks allow for great individual, social, and cultural variance in terms of both their particular features, their concrete performance, and their aesthetic appreciation. As a result, their principal

21 Dissanayake, "The Arts after Darwin," particularly 258.

22 To sketch out four further limitations of Dissanayake's account: (1) She refers to the decorative-visual arts, which Darwin discusses in great detail, only in passing. In general, in her critique of the sexual selection hypothesis, she never cites Darwin's own position, but only Neo-Darwinist variants. (2) The key aspect of aesthetics according to Kant, Darwin, Dawkins, and many others—evaluation, "aesthetic judgment," and the powers of aesthetic preference and choice—is dealt with only in a few remarks on the limited applicability of the predicate "beautiful." (3) Accordingly, the perspective is shifted from the recipient to the "artifying" creator, and hence to the production end that typically was the key focus of poetics. (4) Finally, the significant differences between the individual arts are not an issue. For a more detailed discussion of Dissanayake's understanding of the human arts see Davies, *The Artful Species*, 129–32.

23 Cf. Bourdieu, *Distinction*.

24 Cf. Gans, *Originary Thinking*.

shareability does not translate into equal accessibility and acclaim across times and cultures, and not even across individuals of the same social group. A great share of the respective differences can be explained by familiarity and mere exposure effects,[25] and hence by ontogenetic factors which play in general a great rule in aesthetic liking. Precisely for this reason, artistic traditions can form part of distinctive group-specific cultural traditions, practices, and identities, just like language does. As a result of differential exposure and learning trajectories, understanding artistic traditions and joining in songs and dances can be simple and "cheap" for insiders, but difficult and "costly" for outsiders.

Already simple vocalizations allow for a double categorization: while functioning as "cheap signals" within a group, they can simultaneously be costly for individuals from other groups—if, despite their simplicity, these signals are difficult to copy or simulate. This is the case with local "dialects." The calls of chimpanzees show subtle phonetic variations from group to group, and non-group members cannot readily simulate these distinctive acoustic characteristics of a given group's calls. Members of a group can thus immediately make a categorical distinction between in-group and out-group individuals.[26]

The situation is similar with human languages. From a certain age, dialects or foreign languages can hardly be mastered to a degree that avoids a treacherous accent. The variety of human languages and their dialectal variations thus supports an easy and instantaneous detection of group membership. They are phonetical passports, difficult to fake. Moreover, human verbal language gives rise to additional and new types of "cheap" cooperative signals that are costly to decipher and produce for outsiders. Within a group, even minimalistic expressions can evoke a wealth of shared allusions, including ideological fictions. Individuals belonging to other groups can hardly fully understand these insider semantics, these "conspiratorial whispers,"[27] even if they command the lexicon and syntax of the respective language in general. And in order to be able to produce such compact messages full of hidden allusions in a foreign language, non-native speakers need to invest great amounts of time and energy into linguistic and cultural learning.

25 Cf. Bornstein, "Exposure and Affect"; Reber, Winkielman, and Schwarz, "Effects of Perceptual Fluency on Affective Judgments"; Zajonc, "Attitudinal Effects of Mere Exposure."

26 Cf. Mitani, "Dialects in Wild Chimpanzees"; Crockford, "Wild Chimpanzees Produce Group-Specific Calls."

27 Cf. Krebs and Dawkins, "Animal Signals," 391; and Knight, "Ritual Speech Coevolution," 71.

The same applies to elaborate artistic practices performed at social gatherings: they are relatively easy and "cheap" only for individuals who have gone through repeated exposure and rehearsal—that is, for those who have actually spent much time on learning these practices—but not for those who have not. Hence the shareability of such practices is not indiscriminate and serves two functions at a time: promoting within-group cohesion and between-group distinction. In this dual capacity, the respective practices by no means fully suspend the dichotomy of costly competitive versus cheap cooperative signals. Rather, they call for a differential assignment of the antithetical categories dependent on whether the focus is on the function of the respective practices for intragroup or intergroup communication.

In terms of evolutionary theory, the crucial question arises: Can socially shared *and* costly artistic practices of singing and multimedia performances be conceived as a special evolved adaptation serving intragroup cooperation and cohesion? As to socially shared animal signaling, there is ample evidence that such signaling can reliably be found species-wide and serves special adaptive purposes. Animals that evolved as social hunters at about the same time and in the same ecological niche as humans did, and that frequently share the kill, are a good example: lions, wolfs, and hyenas. All three species show vocal territorial behavior by means of calls that can be heard very far; lions and wolfs can also produce series of such calls in a coordinated fashion.[28] Several nonhuman primates likewise show territorial signaling behavior. For instance, before and during confrontations with adversaries, chimpanzees produce lengthy sequences of "hoot pants" in a coordinated fashion, sometimes in combination with drumming on large tree roots.[29] Human military music, which appears to be found across cultures, may well be a variant of such nonhuman primate behavior.

However, two major differences apply: joint displays of human singing, dancing, and instrumental music are far more varied and hence involve more learning, and they are also far more artistically elaborate than those found in other terrestrial mammals. In Krebs's and Dawkins's typology, the animal counterparts all belong to cheap calls rather than to costly signals. To be sure, the roar of lions and the howling of wolfs may be costly for the animals in that they distinctively inform about their respective physical strength; still, this does not turn these calls into an "art" or into a genuine song of the birdsong type.

---

28 Cf. Hagen and Bryant, "Music and Dance as a Coalition Signaling System"; and Hagen and Hammerstein, "Did Neanderthals and Other Early Humans Sing?"

29 Cf. Arcadi et al., "A Comparison of Buttress Drumming by Male Chimpanzees from Two Populations," 135–39.

In light of these data, one could propose that the human practices of socially shared music making and multimedia performances have a biologically evolved underpinning similar to the one underlying shared animal vocal signaling, and that humans, having biologically evolved to be highly cultural beings, simply add more ontogenetic variance and elaboration to a similar phylogenetic heritage. In any event, this hypothetical add-on results in a categorical difference: only human practices of joint musical and multimedia performances transcend the dichotomy of cheap cooperative versus costly competitive signaling, and this distinctive feature appears to be strongly dependent on the great role of learning and cultural variance.

Contrary to Hagen and Hammerstein,[30] the great majority of the authors who propose that human music making enhances social cooperation and cohesion typically refrain from explicitly claiming a phylogenetically evolved underpinning of this function, be that for reasons of scientific caution or because ethnology and cultural anthropology typically favor an approach to human behavior that focuses on processes of ontogenetic learning and cultural transmission. Moreover, whereas the association of individual artists with great sexual success has haunted human cultural imagination time and again, no analogous myths are found for elaborate joint human music making as serving social coordination. Still, uncertainty and reluctance to engage the difficult question of evolved underpinnings of complex behaviors is far from negative certainty. The issue clearly deserves more empirical research in an evolutionary perspective.

4. The Multiple Blends of Competitive and Cooperative Effects of the Arts

Rites of sexual courtship evolved in order to let some individuals triumph over others. This triumph is exclusive. Social rites, on the other hand, tend to enforce normative expectations, affective codes, and cooperative patterns of behavior which are shared by all members of a respective group. However, at closer inspection, the boundaries between these two types of aesthetically demanding efforts are permeable. Thus, social rites offer many opportunities to accommodate practices of sexual ornamentation. All kinds of body painting and self-embellishment—the classic signals of sexual courtship—reach high levels during ritual events. Individually different abilities for singing and dancing can also be effectively displayed in social

30 Hagen and Hammerstein, "Did Neanderthals and Other Early Humans Sing?"

rites. Hence the stage of confirming a social bond can always be simultaneously used as a stage of individual (sexual) self-presentation. Similarly, practices of self-ornamentation can signal group membership and promote social cohesion among individuals of the same rank, while also highlighting social distance and distinction vis-à-vis individuals from other social strata.

In a similar vein, poetic and rhetorical language use is primarily a competitive art aimed at capturing in a distinct fashion the attention and imagination of an audience and rendering a poem, a narrative, or a speech more moving, more admirable, more memorable, and occasionally also more persuasive than others. At the same time, language has by no means exclusively evolved as a manipulative power instrument but also as a facilitator of social cooperation and coordination.[31] As a result, a narrative that competes with other narratives for the audience's attention and appreciation can end up contributing to the construction of shared cultural horizons and have an all the more cohesive effect, the greater its competitive success and dissemination.

For all these reasons, pointing to the fact that music is mostly practiced in social coordination by no means suffices to invalidate Darwin's theory of competitive sexual proto-music.[32] Rather, the aspects of music that promote sexual courtship and social coordination are by no means mutually exclusive.[33] Exclusively siding with the individual competition *or* the social cohesion hypothesis often has strong ideological implications. In the end, all cultures seem to encompass severe, often violent, conflicts based on divergent individual interests and *also* to provide means of social integration, with different balances of these antagonistic factors in more individualistic and more communal societies. Apparently, special appreciation for individually and competitively "better" arts of painting, singing, dancing, and self-decoration already existed in archaic cultures. Conversely, the highly individualized art production of Western modernity keeps playing an important role for the cultural invention and proliferation of shared horizons of allusions, aesthetic standards, and expectations as well as symbolic evaluations.

The functional poles of individual competition versus group cohesion are all the easier to reconcile as they tend to be respectively associated with art production and reception. On the one hand, musicians, writers,

---

31 Cf. Tomasello, *Origins of Human Cognition.*

32 Mithen (*The Singing Neanderthals*, 180) draws this premature conclusion.

33 Unlike Dutton (*The Art Instinct*, 226), Kirschner and Tomasello ("Joint Music Making," 361) agree with this view.

and painters compete among each other for the attention and favor of the audience, including potential clients and sponsors. On the other hand, their products can provide a shared horizon of aesthetically mediated perceptions, values, and interpretations to the audience. Sharing aesthetic preferences can, in turn, also have a group-dividing implication. Musical preferences that I share with a subgroup of a larger social entity separate me from other subgroups with different tastes; analogously, the interpretation of a myth or artwork that I share with one group separates me from groups who prefer other readings.

As Simmel has shown,[34] similar considerations apply to fashions, styles, and brands of clothing. Among themselves, different styles and brands compete. To their clients, however, they provide shared aesthetic signatures of self-presentation. Despite all emphasis put on individuality and distinction, fashions are also readable as signals of belonging to certain groups sharing certain preferences. The more differentiated our societies, the more selectivity can we expect from the subgroups that constitute such communities of shared aesthetic preferences. These subgroups compete with one another for followers, acceptance/prestige, and social advantages.

Summing up, the competition narrative and the social cohesion narrative of the human arts are likely to apply to different degrees and in varied combinations and hybridizations, depending on historical and cultural circumstances. A general formula granting one factor more weight than the other—or a generalized "both are equally true" account—can hardly capture once and forever the ongoing dynamics of this interaction. In any event, no such combinations of apparently antithetical functions can be found in the animal arts which always and unambiguously lean toward the competitive pole and to exclusive benefits for individuals only.

---

34 Simmel, *Fashion*.

3

Engagement in the Arts as Ontogenetic Self-(Trans-) Formation

Classical theories of "aesthetic" or "liberal education" typically neither consider potential ancient evolutionary origins of the human arts nor entail theoretical comparisons with singing, dancing, and other types of display as practiced by animals. Accordingly, the sexual selection hypothesis does not form part of the prevalent understanding of engagement in the arts in Western modernity. At the same time, as shown earlier,[1] (Western) mythologies of art, poems, and other sources do associate artistic performance and sexual success in a fairly direct way, suggesting that the sexual selection hypothesis advocated by Darwin is by no means an outlier in the broader cultural record.

Compared with the "sexy singer" hypothesis, the social cooperation/cohesion hypothesis has a stronger representation in the humanist tradition; Kant's theory of an aesthetic *sensus communis*[2] is a preeminent example. Still, the predominant hypothesis of function that is associated with the notion of "aesthetic education" does not bear on interpersonal effects (be they of a sexually competitive or a socially affiliative nature), but first and foremost on art-driven *intrapersonal* effects regarding the personal development and well-being of those who engage in the arts. Such effects are henceforth called *self-formative*, or *self-transformative*, benefits. (Giving the humanist hypothesis a more technoid sounding name, Patel

1 Cf. p. 58–60.

2 Kant, *Critique of the Power of Judgment*, 122–24.

relabeled the engagement in the arts as a "transformative technology of the mind," or TTM for short.[3])

Specifically, the hypothesis of ontogenetic self-(trans)formation predicts that engaging in the arts may

- enhance socio-cognitive (theory of mind, mentalizing) and socio-emotional (empathy, compassion) abilities (this hypothesis enjoys particularly great attention in recent research);[4]
- stimulate, exercise, and challenge complex cognitive abilities, such as dealing with ambiguities, indeterminacy, and numerous options of interpretation;[5]
- promote situational and general well-being through providing inherent aesthetic self-reward;
- exercise and improve some motor and technical capacities;
- serve emotional self-regulation (mood management) through selectively fine-tuning one's art reception to one's situational emotional needs;
- enhance school performance in math, reading, and writing;
- prepare us for expectable future life situations and concomitant ethical problems in a simulation mode relieved from the constraints of everyday reality;
- allow us to explore (fictional) alternatives and possibilities of thinking and acting differently.

Neither Darwin nor other authors have considered the various hypotheses of ontogenetic self-transformation as serious contenders for a *phylogenetic* origin of human art-related behavior. In fact, it is hard to imagine how the arts could have evolved in the first place if their functional benefit was *all* about the cognitive, motoric, and affective ontogenetic *self*-formation of individuals qua individuals. Without feedback from others, there would barely be any self-reinforcing reward for special artistic abilities and thus no dynamic development of whatever art, be it self-ornamentation,

---

3 Patel, "Music, Biological Evolution, and the Brain."

4 Cf. Johnson, "Transportation into a Story Increases Empathy, Prosocial Behavior, and Perceptual Bias toward Fearful Expressions"; Kidd and Castano, "Reading Literary Fiction Improves Theory of Mind"; Mar et al., "Bookworms versus Nerds: Exposure to Fiction versus Non-Fiction"; Sopčák et al., *Transdisciplinary Approaches to Literature and Empathy*; and Zunshine, *Why We Read Fiction?*

5 Cf. below, p. 112–15

singing, narration, or painting. *With* social feedback, on the other hand, the presumably purely self-formative practices are actually (also) driven by a genuinely interpersonal and social dynamic, and the competition and the social cooperation narratives are strong contenders that can explain this dynamic.

Assuming that art-driven effects on human ontogenetic development may not be per se informative about the potential evolutionary origins of human art-related behavior, does not imply that such effects are only of secondary importance. In fact, in current environments, transformative effects on the self may well be the strongest functional effects of engagement in the arts altogether. However, because this book is primarily devoted to evolutionary theories of the arts, here the hypothesis of ontogenetic self-transformation is not discussed in its own right and in any detail, but exclusively with regard to the following question: How compatible is this hypothesis with the sexual competition/social distinction and the social cooperation/cohesion hypothesis?

It is not difficult to identify reasons for why the hypothesis of self-transformation by engaging in the arts is, in principle, not incompatible with the functions proposed by the sexual competition/social distinction and the social cooperation/cohesion hypotheses:

1. If the human (musical) arts evolved as competitive practices of sexual courtship, then practices of training and refining these abilities should have evolved in humans no less than in birds. These practices would almost by definition involve some ontogenetic effects upon those who practice them. Accordingly, Darwin readily embraced the notion that "the habitual practice of each . . . art must . . . in some slight degree strengthen the intellect."[6] Obviously, then, Darwin was not concerned regarding the compatibility of his evolutionary theory of aesthetic evaluation and the arts on the one hand and ontogenetic effects of engagement in the arts on the other. In fact, as already noted above, the very reverse applies: he exclusively speaks of "art"—even regarding animals—where learning and exercise is required and hence some ontogenetic refinement of individual capabilities is involved.
2. If the human (musical) arts evolved as practices of social cooperation/cohesion, a similar consideration applies. Under this premise, too,

6 Darwin, *Descent of Man*, II 161.

and for the very same reason, practices of training and refining one's performance should have arisen (even though the degree of competitiveness should be lower). Again, these practices should have yielded formative effects on the practitioners. As a result, art-driven self-formative effects can readily be conceptualized as an implication or a side effect of both the sexual selection and the social cohesion model.

3. Inversely, if engagement in the arts in fact evolved—by means of whatever mechanism not treated in this book nor in any other evolutionary account of the arts to date—as cultural techniques of self-formation in the first place, then the potential beneficial effects on the personal development of individuals should in the end also affect their social life. The more valued a particular art is, the more can good performance in this art also be a criterion for competitive social esteem. Moreover, individuals who excel in artistic capacities that enjoy social esteem may also be better off in terms of securing sexual partners. Finally, once established as self-formative techniques of individuals, some artistic practices could have gained more social acceptance than others and in the end become cultural fashions with a collective signature. As a result, self-formative practices could also end up contributing to defining cultural identities of groups and hence to the social bonds within the respective groups.

Thus, the three functions of the human arts considered here can well positively correlate with one another and therefore are not mutually exclusive. This is in good accord with the notion that the human being is both a very social *and* a very competitive being, and moreover one that is more dependent on ontogenetic learning than presumably all other species. Responses to the question "Wherefore art?" should therefore be careful to avoid false alternatives.

4

A Cooptation Model of the Evolution of the Human Arts: When "Sense of Beauty," Play Behavior, Technology, and Symbolic Cognition Join Forces

According to the standard model of evolution, novel—or altered—traits of an organism evolve as *domain-* and *task-specific* adaptations. Darwin's account of ornamental body traits and artistic displays in animals—and of a correlative "sense of beauty" driving these traits—largely conforms to this assumption. However, both Darwin's and more recent concepts of evolution also entail provisions for other types of evolutionary processes. Once evolved, all given task-specific adaptations as well as nonadaptive side effects of these adaptations constitute a "pool of features available for cooptation."[1] This implies that it may not always be necessary for evolution to modify a given trait or add a task-specific novel one. Rather, the emergence

1 Gould and Vrba, "Exaptation," 13. Cf. also Fitch, *The Evolution of Language*, 63–66. Gould and Vrba call such adjustments "exaptations." However, this term has some counter-intuitive implications. The prefix "ex-" overstresses the distancing from an existing adaptive function; by its very linguistic nature, it makes it seem rather improbable that these "exaptations" can be new "adaptations." A spontaneous understanding of this neologism also leaves one wondering why the adaptive cooptation of an existing *non*-adaptive feature is supposed to also qualify as an "exaptation." Here, one could rather expect the term "enadaptation." For these reasons, the term "exaptation" will not be used

of a new behavior can just as well rely on "coopting" in a novel way existing traits for novel functions, without necessarily altering, let alone abandoning, their original task-specific function.

As Gould and Vrba point out,[2] already Darwin gave examples for such cooptations, specifically with regard to complex traits of human behavior.[3] The present chapter proposes a cooptation hypothesis that explains how humans extended the range and functions of the arts far beyond the bird arts and other animal arts. They did so, according to this hypothesis, by means of coopting their adaptations for play behavior, tool use, and symbolic cognition/language—all of which developed independently from the human arts, and at least the first two of which are likely to be (far) older than the arts—into the field of art production and perception/evaluation while extending evaluations for beauty far beyond sexual bodily beauty.

1. The "Sense of Beauty"

The "sense of beauty" as conceptualized by Charles Darwin[4] keeps being a crucial backbone of the cooptation model of the human arts presented in the following pages. After all, in an evolutionary perspective, the perception and evaluation of artworks is no less important than the capacities of producing them. As discussed in Chapter 1, Darwin sees evidence for sexual choice in humans that is driven by the natural beauty of looks, artificial enhancements of these looks by self-made ornaments, and beautiful singing. At the same time, much of Darwin's panorama of beauty-related practices of self-ornamentation and self-shaping is about interactions of a hypothetically biologically evolved mechanism of sexual self-presentation and choice with a great variety of culturally varying and ontogenetically acquired practices. Moreover, even though the modern arts partly question the primacy

in this book, even though its argument relies on Gould and Vrba's insights. Instead, I shall exclusively talk of new uses and new cooptations of old evolved traits.

For similar reasons, this study likewise does not adopt the term "pre-adaptation." When existing traits support the emergence of future (later) adaptations, they are occasionally called "pre-adaptations" (cf. Wilson, *Sociobiology*, 34, and Gould, "Adaptation," 48). This term too, provokes somewhat misleading associations. Importantly, an old adaptation can be coopted and used in novel ways, without this potential resulting from any kind of anticipated adaptation (let alone pointing to some prescient teleology of a plan underlying nature).

2 Gould and Vrba, "Exaptation," 5 sqq.

3 See, for instance, Darwin, *Descent of Man*, II 335.

4 Darwin, *Descent of Man*, I 63–64.

of striving for beauty, empirical studies suggest that beauty keeps being the first and foremost though by no means the only aesthetic virtue that guides art-related expectations on the part of the wider audiences.[5]

Combining behavioral and neuroscientific methods, two recent studies[6] could show that aesthetic evaluation routines, specifically beauty evaluations, are virtually always and automatically "on," if humans look at faces or read fairly inconspicuous sentences, even as they lie in the highly alienated environment of a noisy MRT-scanner and are actually deviated from any conscious focus on aesthetic evaluation by means of other tasks they are asked to fulfill.

2. Sexual Courtship, Play, and the Arts

Sexual courtship and choice are not playful and autonomous, at least not in the animal kingdom. Peacocks and other birds must demonstrate their bodily, musical, dancing, or building capacities—or perish (at least, with regard to the proliferation of their genes). When partner choice is based on aesthetic competition, artful displays are a very serious business. The effort to cut a good figure in this competition is unconditional and "existential" in a very basic meaning. Accordingly, the presenters are invariably extremely devoted to their efforts. The image of the human artist as driven by his artistic project fits in well with this very strong drive for aesthetically elaborate display observable in sexually courting animals.

A comparative glance at nonhuman primates and other mammals confirms this distance from play. From an evolutionary perspective, playing primarily exercises behaviors such as hunting, self-defense, fleeing, and attacking; it has little or nothing to do with dancing or singing displays for the other sex. Many birds do *exercise* their vocal abilities, but ornithologists do not regard this behavior as play. On a more general note, play behavior is barely ever connected to sexual selection in evolutionary biology. Rather, species-wide dispositions for play behavior are typically explained with the standard model of natural selection.[7]

---

5 Cf. Knoop et al., "Mapping the Aesthetic Space of Literature 'from Below'"; Augustin et al., "Artful Terms: A Study on Aesthetic Word Usage for Visual Art versus Film and Music"; Augustin et al., "All Is Beautiful? Generality vs. Specificity of Word Usage in Visual Aesthetics"; Istók et al., "Aesthetic Responses to Music."

6 Chatterjee et al., "The Neural Response to Facial Attractiveness"; Bohrn et al., "When We Like What We Know."

7 For a more complex view, see Burghardt, *The Genesis of Animal Play*.

This is very different for the human arts. They can be—and have been—connected both to Darwin's model of sexual selection *and* to play behavior.[8] Both approaches seem well justified. Typically, they are pursued independently from one another. Therefore, the (in)compatibility between the two explanations has barely been discussed: How can the human arts both have an evolutionary antecedent in the "serious" business of sexual selection and be a form of play behavior? Darwin's observations on the (seemingly) functionless pleasure that many birds take in singing outside the courtship season offer a potential link between highly serious sexual courtship and playfully practicing one's singing and other capacities:

> It has been argued, that the song of the male cannot serve as a [sexual] charm, because the males of certain species, for instance of the robin, sing during the autumn. But nothing is more common than for animals to take pleasure in practicing whatever instinct they follow at other times for some real good. How often do we see birds which fly easily, gliding and sailing through the air obviously for pleasure? . . . Hence it is not at all surprising that male birds should continue singing for their own amusement after the season for courtship is over.[9]

In this view, animals, too, can practice the arts of sexual courtship as wholly self-rewarding and play-like activities without direct pragmatic benefits ("real good"). Recent research has shown that birdsong shows markedly different neural underpinnings dependent on this difference in context (singing for females vs. singing in the absence of females).[10] Darwin reports extreme cases of obsessive singing:

> That the habit of singing is sometimes quite independent of love is clear, for a sterile, hybrid canary-bird has been described as singing whilst viewing itself in a mirror, and then dashing at its own image [. . . For a bird-catcher,] the test of a really good singer is that it will continue to sing whilst the cage is swung round the owner's head.[11]

The analogy to hard forms of artistic "madness" is obvious. In any event, musical displays can be associated both with serious intra-sexual competition and with self-rewarding pleasure in the absence of any immediate

8 Cf. Boyd, *On the Origin of Stories.*
9 Darwin, *Descent of Man*, II 54.
10 Jarvis et al., "For Whom the Bird Sings."
11 Darwin, *Descent of Man*, II 53.

sexual goal: "Many birds . . . delight in their own music, and try to excel each other."[12] In all likelihood, pleasure taken in singing for the sake of singing and a propensity to engage in musical competition even in the absence of pragmatic goals improve performance skills through exercise and thus eventually contribute to the success in sexually competitive singing. Playful, purposeless practicing of an art and functional benefits for the future life of an organism are not mutually exclusive, be it here or elsewhere. Only in this sense can the model of sexual singing be reconciled with the topical modern notion that the arts are largely driven by playful exploration and purely self-motivated elaboration.

Play behavior is prominent in many mammals and birds,[13] including nonhuman primates. Apart from play hunting and play fighting, apes and monkeys also show play behavior that is explorative and apparently guided by cognitive curiosity.[14] This kind of play resembles the corresponding play varieties of human children. In combination with the capacity for tool use, play behavior might be the evolved basis for such phenomena as the chimpanzees' exploration of previously unknown behavioral opportunities that open up when humans provide them with paper and painting tools.[15] Accordingly, Bernhard Rensch entitled his meta-analysis of a large number of studies on this intriguing issue "Play and Art in Monkeys and Apes."

It is widely agreed on that play behavior serves to promote the acquisition and improvement of vital communicative and motor capacities, particularly the competitive capacities for hunting, attacking, and self-defense. The playing parties are usually not—or at least do not need to be—aware of these and other medium- and long-term functional benefits of their behavior. Playing is far more likely to be experienced as purposeless, as driven by a self-motivating pleasure of playing for the sake of playing, fully immerged in the here and now. Topical comparisons of play and aesthetic pleasure imply this notion, too.

In play, the decoupling of relevant action patterns from their real-life implications provides a largely risk-free (learning) environment,[16] a parallel

12 Darwin, *Descent of Man*, II 277.

13 Cf. Fagen, *Animal Play Behavior*; Panksepp, "The Ontogeny of Play in Rats"; Panksepp et al., "The Psychobiology of Play: Theoretical and Methodological Perspectives."

14 Rensch, "Play and Art in Apes and Monkeys," 105–10.

15 Cf. Rensch, "Malversuche mit Affen"; Morris, *The Biology of Art*; Lenain, "Ape-Painting and the Problem of the Origin of Art," 207–9.

16 Tooby and Cosmides, "Does Beauty Build Adapted Minds?"; Storey, *Mimesis and the Human Animal*; Deacon, "The Aesthetic Faculty," 30 sqq.; Eibl, *Kultur als Zwischenwelt*, 164–69; Neumann, *Funktionshistorische Anthropologie der ästhethischen Produktivität*, 36–61.

world (a "twin earth") that exists alongside the real one.[17] The different worlds have different cognitive frameworks that code for different evaluations and consequences of near-identical actions. Rather than leading to cognitive confusion, precisely this ontological split is of vital relevance for survival. In the end, occasionally adopting an ex-centric position of suspended (ordinary) reality enables players to better cope with this reality. Human play behavior, which expands this general potential of play by virtue of also drawing on the distinctively human capacities of language and symbol use, is a great school for all techniques of fictional representation and simulation, including artistic ones.

While priming an ongoing awareness that one is not engaged in ordinary reality, cognitive mind frames or schemata[18] of play and art production/reception do not preclude states of deep immersion (engagement with games, music, a book, or a movie, in which one is "dead to the world"). Rather, the two poles apparently can be combined in different proportions, and even a deeply immersed reader hardly ever fully loses track of the ontological difference. To be sure, from Cervantes's *Don Quixote* to Woody Allen's *The Purple Rose of Cairo*, works of art have staged humoresque scenarios in which the boundary between text or screen and reality is, in fact, completely disregarded by protagonists of artworks. However, this prevents recipients all the more from being subject to the same deceit.

The arts multiply the play with different cognitive frameworks, each of which encodes a world of its own with distinctive rules. Every genre, indeed every single artwork, requires and implements the implicit construction of its own "playing field" with a special set of rules. In this regard, playful and artistic practices resemble axiomatic systems. By creating numerous and highly self-referential worlds, works of art question naïve "*one* reality" realism not only through the binary opposition of play versus reality but also through subtle nuances distinctive of their respective anti- or para-realistic ontology. Thus, they open our eyes to the selectivity of our dispositions for action and perception and get us accustomed to the idea that everything could be somewhat, or completely, different. Artistic practices thus entail a meta-reflective relationship to established habits of perception and behavior; they encourage the opening of options for new variation. For this very reason, modern poetics and aesthetics tend to emphasize, indeed

17 Lillard, "Pretend Play as Twin Earth." Cf. also Cosmides and Tooby, "Consider the Source: The Evolution of Adaptations for Decoupling and Metarepresentation."

18 Cf. Abelson, "Psychological Status of the Script Concept"; and Brewer and Nakamura, "The Nature and Functions of Schemas."

to overemphasize, a close connection between art, de-automatization, estrangement, innovation, and transgression.[19]

Cognitively switching from a reality into a play mode provides opportunities for training and (self-)regulating not only motor patterns but also affective responses. Aggressive and defensive affects are equally drawn on in almost all forms of play. Pleasure and disappointment about successful and unsuccessful moves frequently take turns in quick succession. If contingency, or "fortune," would not limit the predictability of the results, games would hardly be games. Playing thus presents a double challenge: first, for one's intellectual and motor capacities, and secondly, for the affective handling of scenarios in which opposite developments—good and bad luck, successes and failures—are always possible. This is why Kant considers the anticipation of the next turn in games of chance an alternation between "fear and hope," "perplexity" and "joy."[20] Without the chance of being disappointed or even annoyed at every new turn, the joy about one's good luck or the others' bad luck would not have any greater amplitude and resonance.

Thus, in play, negative and positive feelings function as communicating vessels. Kant's reflections[21] about the "animating" or "enlivening" (*belebend*) effects of games and art reception draw on a theoretical concept that plays a crucial role in eighteenth-century aesthetics, also belongs to Fechner's basic "principles" of aesthetics[22] and was recently rediscovered as an explanation of the "pleasures of the mind":[23] the concept of a dynamically changing interplay, or interaction, of a variety of emotions. This concept involves the assumption that single positive or negative feelings—be it love, admiration, or sadness—are an insufficient basis for a sustained and self-reinforcing interest in works of art. As all acutely felt emotional episodes are limited in duration, a continuous orgy of admiration would become boring just as quickly as a painting that only presents beautiful things in a beautiful way. Moreover, the demand for alternation and emotional change corresponds to the elementary experience that social communication nearly invariably tends to lead to conflicts at some point, and hence can barely be represented without recourse to negative emotions. Thus, the affect-poetic program that calls for artful interplays of opposite emotions reflects both the inherent instability of single emotional states, the conflict potential of social

---

19 Shklovsky, "Art as Device."

20 Kant, *Anthropology from a Pragmatic Point of View*, 127.

21 Kant, *Critique of the Power of Judgment*, 208. See also Menninghaus, "Ein Gefühl der Beförderung des Lebens."

22 Fechner, *Vorschule der Aesthetik*, 246–53.

23 Cf. Kubovy, "On the Pleasures of the Mind," 134–54.

communication, and the unique powers of emotional change to capture and bind an audience's attention.[24] The hypothesis of emotional changes, or blends, also corresponds to the psychological hypothesis of cognitive contrast, according to which we value things all the more if their positive aspects are highlighted by a negative contrast.[25]

Given the many psychological mechanisms shared by play and art production/reception, theories of play and art typically converge in the key assumption that, basically, both play and the arts are about *inherent processing pleasure* detached from direct pragmatic concerns. This holds true regardless of the fact that some artworks can well be challenging, shocking, disturbing, and may not be pleasant in a narrower meaning, just as some plays can be—physically or intellectually—very demanding. Moreover, it is widely assumed that playing and art reception provide functional benefits beyond the immediate processing pleasure. On the one hand, they may already do so, because experiencing inherent and "enlivening" processing pleasure can enhance mood states past the end of actually playing or seeing a movie. On the other hand, playing and art reception/production provide our nerves and muscles, as well as our cognitive and affective dispositions, with opportunities and materials for pleasurable activation. As all our highly flexible capacities depend on usage ("use it or lose it"), play and art reception are in this regard also inherently advantageous from a functional point of view.[26]

One of the basic prerequisites of human engagement in representational artworks is that we can be emotionally affected by the depiction of the happiness and suffering not only of absent or dead but also of purely fictitious or "performed" persons, and even of clearly fantastic beings. Pretend play, in which children engage from about eighteen months of age onward, shows that this "mad" capacity for genuine emotional involvement in fictional beings and worlds has a basis in early forms of symbolic play—such as takes place when, for instance, a child defines a piece of wood as a person in order to then use it for staging social dramas of great affective salience.[27] Interestingly, pretend play, verbal language, and a general capacity for hypothetical representation develop nearly simultaneously at the same time

24 For more details on this theory, see Menninghaus et al., "The Distancing-Embracing Model of the Enjoyment of Negative Emotions in Art-Reception."

25 Cf. Parker et al., "Positive and Negative Hedonic Contrast."

26 Martinelli, "Liars, Players, and Artists," 96. Cf. Nottebohm, "From Bird Song to Neurogenesis," 74–79.

27 For more details on engagement in fictional agency and narratives, see pp. 104–112.

in human ontogenesis.[28] Moreover, whereas play and artistic displays are clearly separated in the ontogenesis of animals—with play behavior largely ending before sexual maturation and artistic displays being confined to the sexually most active period of adult life—humans engage in both play and art reception/production well into old age.

Taken together, the similarities between human play behavior and the arts identified so far—the analogous cognitive frameworks; the licenses, experiences and chances to learn in the "offline" mode of suspended reality; the animating, "enlivening" effect of the sheer activation of our senses and affects; the unusual and substantially overlapping extension of play behavior and engagement in the arts throughout human ontogenesis—strongly suggest that the human arts profited from the presumably much older practice of play.

At the same time, there is a marked difference between the two. Play typically does not produce a lasting material product. Rather, the energy, pleasure, and benefits of playing primarily refer to the very practicing of the play capacities themselves. Artistic behavior, in contrast, often does produce lasting material objects. The performative arts, however, are as ephemerous as play; they, too, leave no material traces beyond their sheer performance, at least if not somehow recorded.

3. Technology and the Arts

Darwin's theory of sexual selection based on looks and/or excellence in singing, dancing, and even "architecture" (such as the building of nests and bowers) suggests four pretechnological paths to the human arts: (1) a visual one, via the sensitivity for major and minor differences in physical attractiveness (natural body "ornaments"); (2) a further visual one, via the sensitivity for excellence in self-displays by means of body movements (dancing) or in building decorative structures; (3) an auditory one, via the sensitivity for different song quality; and (4) an audiovisual one, via the sensitivity for the aesthetic appeal of visual displays of plumage and/or movements that are combined with sound displays.

None of these forms of aesthetic display in animals imply tool use. To the extent that the respective species generally do not engage in tool use, this

28 Meltzoff, "Towards a Developmental Cognitive Science," 23 sqq. On the evolutionary dissemination of imaginative pretend play, cf. Mitchell, "Pretending and Imagination in Animals and Children"; Lyn, "The Development of Representational Play in Chimpanzees and Bonobos"; Jensvold and Fouts, "Imaginary Play in Chimpanzees."

finding might seem trivial. However, even animals that do use tools hardly ever do so in the context—and much less for the purpose—of aesthetically elaborate courtship practices. Searching for, acquiring, conditioning, and taking in food: these are the typical evolved functions of animal tool use. Going beyond this narrow range, humans use tools also as weapons both for killing animals[29] and in fights with other humans; this extension of tool use is probably much older than the use of tools for purposes of self-ornamentation. To be sure, bowerbirds use their beaks as instruments to build decorative structures for the purpose of sexual courtship; however, this is merely one of many forms in which an animal may use its own body parts as instruments. Such uses are markedly different from the purposeful use of objects *external* to the body, be those found or self-made. With regard to the objects which bowerbirds find and use to decorate their bowers: these are used as a means of courtship, but not as tools in any narrower sense.

For our closest relatives, interesting borderline cases have been reported. Chimpanzees have been observed to swing branches horizontally in order to attract attention and impress the opposite sex in the context of sexual courtship behavior.[30] Another intriguing example is the leaf-clipping display, in which large firm leaves are taken to pieces in a markedly rhythmical movement pattern accompanied by sounds. The display is apparently used in several functional contexts, including invitations to mate. Here, too, naturally found, but also partly processed objects are employed as (expressive) communicative devices.[31] Can we regard this as tool use for sexual and other purposes, even though the object used does not mechanically touch and change any other object? Is this display perhaps even analogous to the use of magic and religious objects believed to have purely mental long-distance effects?

The context of sexual display speaks against interpreting this behavior as tool use in any narrower meaning.[32] Instead, it can rather be conceived as expressive and evocative signaling with the help of an object: the swinging of a branch enhances the signal language of bodily arousal (and adds an acoustic stimulus), and the leaf-clipping has analogous functions. On the

29 Tool-assisted hunting has been observed for a chimpanzee population in Senegal, but to date in no other population of nonhuman primates. See Pruetz et al., "New Evidence on the Tool-Assisted Hunting Exhibited by Chimpanzees."

30 Nishida, "Sexual Behavior of Adult Male Chimpanzees," 385.

31 Nishida, "The Leaf-Clipping Display," 117–28.

32 Unambiguous examples of tool use for the purposes of aesthetic-explorative play with colors and forms could only be observed in the already cited painting experiments with nonhuman primates (i.e., in contexts in which behavior was strongly influenced by human interventions).

other hand, both branch swinging and leaf clipping could be interpreted as particularly advanced forms of tool use, namely, as the equivalent of using choreographic utensils or musical instruments. However, to fully justify this parallel to dancing and music in the sense of Darwin's theory of sexual display, it would need to be shown not only that chimpanzees at least partly acquire these practices through learning (which is likely to apply), but also that the performance involves individual artistic differences that are predictive of the responses from the part of conspecifics.

The hypothetical human proto-arts, which Darwin sketched in direct analogy to the singing and dancing of birds, do not involve tool-based technologies. In contrast, the oldest human artworks for which we have definite archeological evidence (i.e., pictures, statues, and decorative objects) do require *technologies*. Musical instruments preserved from the time some 30,000 to 40,000 years ago likewise involved tool use, suggesting that this link between technologies and the arts of singing and music making dates back a very long time. These well-documented artistic practices, above all Paleolithic ornaments and visual artworks, thus feature two important and distinctive characteristics: they result in (potentially) lasting works of art (rather than being confined to materially traceless performances, i.e., acts of singing and dancing), and they involve tool use (technology). In the case of music, however, these material technology-based traces are limited to the musical instruments and do not include the artworks themselves. The absence of these two characteristics—the object character of the artwork and dependence on tool use—in the hypothetically far older human practices of tool-less singing and dancing explains our lack of empirical knowledge regarding these practices.

Ethnology provides a wealth of evidence for Darwin's assumption that practices of singing and dancing are widespread in archaic cultures. (Still, it would be risky to equate these recent archaic cultures with those of the Paleolithic.) Moreover, ethnologically documented artistic practices almost always involve technology: musical instruments appear as frequently as technically created ornaments, masks, etc. Thus, these practices, too, support the hypothesis that the human arts (of the past 40,000 years at least) call for being conceptualized in conjunction with a major advance in technological diversity (i.e., with special technologies being developed for an increasing array of special practices far beyond hunting, fighting, food preparation, and clothing).[33] Darwin's secondhand reports on elaborate aesthetic practices in African, Asian, and American peoples are well in line

33 Cf. Gell, "The Technology of Enchantment and the Enchantment of Technology."

with this assumption.[34] They are primarily about technically created or at least technically processed ornaments of all kinds as well as about spectacular technologies of purposefully deforming the human body in the service of cultural ideals of beauty that often seem bizarre to a Western reader.

Tool usage appears to be largely limited to the higher mammals.[35] In recent decades, an increasing number of examples of animal tool use—and of locally different tool uses within the same species—has been observed. Correspondingly, many biologists speak of learned and transmitted "cultures" in animals. Chimpanzees are among the tool-using apes.[36] Several great apes also purposefully create tools by modifying found objects to a lesser or greater degree.[37] This suggests that hominids may have had a capacity for tool use from the very beginning. This assumption is also supported by evidence that the neural signatures of tool use in humans can be interpreted as a further development of a similar processing pattern in apes.[38]

According to the contemporary state of research, a new characteristic of human technology is the systematic use of tools to create other tools.[39] This second-order tool use enabled humans to make instruments from materials far harder than those used by nonhuman primates: particularly, stone and mineral ores. The combination of second-order tool use and very hard materials has great archeological consequences. Only this combination could produce large amounts of datable technological evidence from a very distant past.[40] According to conservative data, archaic humans have been making and using stone tools reliably and frequently for at least

34 Darwin, *Descent of Man*, II 338–54.

35 There are also reports of tool use in a few other species, such as crows and ravens.

36 Cf. Dunbar, *The Human Story*, 145–55.

37 Boesch, "Tool Use and Tool Making in Wild Chimpanzees," 86–89; van Schaik, "Manufacture and Use of Tools in Wild Sumatran Orang-Utans," 186–88.

38 Cf. Culham and Valyear, "Human Parietal Cortex in Action"; and Dehaene and Cohen, "Cultural Recycling of Cortical Maps," 393.

39 Cf. Mithen, *The Prehistory of the Mind*, 106–8. However, such second-level tool use has occasionally been observed in apes, too, at least in individuals living in captivity. Cf. Lethmate, "Tool-Using Skills of Orang-Utans."

40 Animal-made tools from organic materials can weather such long periods only under extremely exceptional circumstances. Softer tools usually disintegrate much too quickly, and harder objects that might have been tools used by animals cannot be identified as tools with enough certainty because they were not (or hardly) modified for use as tools and/or have not been used in the same function sufficiently often to leave clear-cut traces of use. The latter problem usually also applies to stones that are sometimes used as tools by apes.

two million years. Recent studies even suggest that this state of affairs is as old as 3.4 million years.[41]

Human bipedalism—whose evolution, about four million years ago, precedes the oldest clear evidence for self-made tools[42]—liberated human hands from their involvement in locomotion and thus promoted the development of tool use.[43] However, the progress in the differentiation of tools as documented in the archeological record has been very slow up to the latest 500,000 years or so. This is all the more surprising as it is precisely in this long phase of apparent technological stagnation that our human ancestors experienced a dramatic increase—in fact a doubling—of their brain volume. Accordingly, the social brain hypothesis states that this burst of intellectual development in humans was less driven by demands for increasingly complex tool-based technologies than by heightened demands on the processing of social complexity.[44]

Regardless of the apparent absence of major technological advances over more than a million years, humans did have a significant competitive advantage over other species throughout this period: even their relatively simple multifunctional stone tools appear to have been quite effective. They were not easy to make, either. Even after lengthy exercise, contemporary university students enrolled in technical courses and instructed in the hypothetical archaic technology of producing tools regularly fail to produce similar stone tools that are of acceptable quality.[45] Archaic stone tools thus were valuable on three counts: they were useful for enhanced performance in a variety of specific tasks; they reflected a significant amount of invested time and energy; and they proved that the toolmaker was in command of the necessary skills.

These considerations suggest that tools played an important role in the social and affective economy of human groups. *If humans were hunter-gatherers for by far the longest part of their history and if they primarily used self-made tools, then the critical evaluation of tools and the capacities of creating them must have been essential for survival and thus relevant to natural selection.* If so, good technical capacities probably also

---

41 McPherron et al., "Evidence for Stone-Tool-Assisted Consumption of Animal Tissues before 3.39 Million Years Ago."

42 Cf. Dunbar, *The Human Story*, 22 sqq., 30.

43 Darwin, *Descent of Man*, II 141.

44 Cf. Humphrey, "The Social Function of Intellect," 303–17; Dunbar, "Coevolution of Neocortical Size, Group Size and Language in Humans," 681–735; Dunbar, "Grooming, Gossip and the Evolution of Language"; Dunbar, "The Social Brain Hypothesis"; Dunbar, "Evolution in the Social Brain," 1344; Kummer, *The Social Intelligence Hypothesis*.

45 Cf. Mithen, *The Prehistory of the Mind*, 134–36.

created social recognition.[46] In this context, the aesthetic evaluation of tools regarding their symmetry, precision, and finish—and occasionally also their ornamentation—could provide important information about the relevant capacities and arguably also about the social rank of individuals. Thus, tool quality could selectively motivate decisions as to who might be a particularly valuable partner for cooperation and social bonding. Archaic practices of tool evaluation, if gradually transferred onto other object areas, might even have paved the way for the broader range of aesthetic judgments regarding cultural artifacts found in more modern humans.[47]

The "sexy hand axe" hypothesis pushes this reasoning to an extreme: it reduces the well-founded importance of tool aesthetics into just another variant of sexual choice.[48] Referring to the existence of many well-preserved, almost unused hand axes that are more symmetrical than practical use seems to require and, moreover, often as big as to be unwieldly, this hypothesis stipulates that men of the early Paleolithic period (and thus long before the times of Homo sapiens) literally had to display their capacity for producing hand axes in the immediate context of sexual selection. According to this view, only producing well-made hand axes before the eyes of interested observers could ensure that these axes were not inherited, obtained in a fight, or stolen, but actually made by the respective individual. This display practice supposedly led to an overproduction of axes that were, in fact, not needed, thus explaining the many Paleolithic axes preserved in surprisingly "mint" condition. The hypothesis fits in well with the notion of unnecessary expenditure ("waste") in costly signal theory. The courtship behavior of bowerbirds could be referred to as an analogous display behavior. Male individuals of these species must construct large, well-proportioned, garishly decorated, but otherwise unused bowers, and female birds apparently base sexual choice on critically evaluating these bowers for their design and aesthetic finish[49]. Similarly, the lesser or greater symmetry of hand axes could have decided the sexual success of early paleolithic men.

Alas, this witty conjecture is hardly compatible with the known courtship rites of Homo sapiens sapiens. It is also hard to see how it could do justice to the complex strategies of finding and evaluating partners in social groups with diverse hierarchies and unstable alliances. The peacock and bowerbird model of partner choice has completely different social

46 See Wilson, "On Art," 72; similar in Dutton, *The Art Instinct*, 175, 191 sqq.

47 Washburn, "Comment."

48 Cf. Kohn, "Handaxes: Products of Sexual Selection?," 518–26; Mithen, "The Singing Neanderthals," 189–91; and Miller, "The Mating Mind," 288–91.

49 For a different explanation, see Prum, 199–205.

preconditions; it would require substantial extra efforts of explaining why this model should work similarly in a highly social species.

Nevertheless, there is strong evidence for the social importance of well-made, and also beautiful, tools in domains other than direct courtship displays. The high affinity between (male) ornaments and weapons—observable in any ethnological museum, extolled by Homer and deplored by Sappho[50]—is the cardinal example here. Shields, swords, daggers, rifles, revolvers: all classic weapons worn at the body have been adorned to highly extravagant and costly degrees and constitute an entire domain of an aesthetics of weaponry, including weapons of defense no less than those of attack. The aesthetic codes of these weapons correlate—or at least used to correlate in earlier times—with signals of social status and prestige. However, the technical and artistic capacities of the wearer himself are hardly ever involved, let alone at stake here. Possessing such weapons is sufficiently reliable proof of material resources and social rank, irrespective of who produced them.

Unlike the feathers, bowers, songs, and dances on which Darwin based his model of sexual choice, weapons and tools always have a technical and practical value apart from the aesthetic one. Therefore, they can and should be set apart from objects of *purely* ornamental importance. Spencer, Darwin's contemporary, had a very precise sense of the type of object that could be seen as a direct equivalent of the impractical peacock feathers in the domain of human ornamentation practices.[51] According to his hypothesis, the oldest ornamental objects were *trophies*. Indeed, archeologists and ethnologists report countless ornaments made from teeth, feathers, bones, and sometimes also skins of slain animals. Interestingly, Stone Age hunters preferred to make ornaments not from the teeth of animals that were probably killed most often (such as reindeer) but from mammoth, cave bear, wolf, and fox teeth.[52] Analogously, other ornamental materials were also subject to the requirement of rarity and costliness: they were only rarely found in the habitat and/or had to be brought from afar.[53] Ornaments were thus literally a "costly signal," one that required an above-average amount of time, hunting skills, and resources. Insofar as ornaments partly

50 Sappho 27a D.

51 Spencer, *Ceremonial Institutions*, 174–92.

52 Cf. Kölbl, "Ich, wir und die anderen," 167. However, it should be noted that, apart from their conformity to the aesthetic principle of costly signals as well as their supposed social and magic symbolic power, the teeth of predators also have a practical advantage for the production of ornaments: they are more durable than ruminants' teeth.

53 Ibid., 169.

derive their value from their trophy character, self-decorative practices fit in well with the fitness indicator theory, and hence with mechanisms of natural selection.

Spencer convincingly argues that military decorations seamlessly continue archaic practices of ornamenting oneself with trophies. After all, badges and medals are pinned on uniforms for special achievements in combat. Often, such decorations are conspicuously colorful. Thus, the concept of a "highly decorated" officer combines the marked decorative appeal with a discreet reference to a pragmatic meaning of this decoration, which essentially lies in the number of enemies killed. Even today, soldiers in dress uniforms appear rather flamboyant and peacock-like compared to an average civilian. Moreover, the domain of *non*-martial ornaments could be understood to be a derivative from the use of headdresses, amulets, pins, and rings as trophies in military contexts. The metaphor "war paint" as used for an intensely made-up female appearance could be speculatively interpreted as a memory trace indicative of this hypothetical martial origin of ornamentation.

Unlike decorative weapons, necklaces and bracelets made from shells or bear teeth—as well as colorful military decorations—have no direct practical value. Their archaic forms (holes in marine gastropods suggest that these were used as ornaments as early as 165,000 years ago)[54] imply perforation, concatenation, and attachment techniques subtle and reliable enough to allow for the transformation of natural bodies (shells) or body parts (teeth) into ornaments enhancing the appearance of another body. The materials that were strung together for this purpose were not yet themselves technically produced, but natural in origin. Still, their configuration and concatenation clearly amounts to self-made and intentionally *designed* ornaments.

Archeological data for ornamental objects that are manmade to an even higher degree only exists for the period beginning with the "creative explosion" about 40,000 years ago. These combine the characteristics of weapons (intense technological investment, three-dimensional objects of a form that is fully defined by human design) with those of ornaments made from found or captured bodies/body parts (purely ornamental use as a ring, bracelet, necklace, brooch, etc.; no practical utility beyond the ornamental, religious, or socio-symbolic contexts). These more recent types of objects are pieces of jewelry made of *technologically processed* rather than natural materials (metals, ivory beads, and others).

54 Cf. Haidle, "Wege zur Kunst," 242.

The cognitive-affective heritage sedimented in our long history of technical tool use seems to have influenced the way we respond to works of art.[55] The appreciation of what is technically "well-made" is an intellectual evaluation of form and function—and simultaneously an affective evaluation that reflects the potential salience of such well-made objects for own functional goals. Among the motivational consequences of such judgments are wishes and inclinations to use, to possess, or at least to seek repeated exposure to the well-made object. Such appreciation for well-made objects shares many characteristics with the preference for certain external features of natural bodies in the domain of sexual choice: these, too, are evaluated not only as beautiful in themselves but also as desirable for and conducive to one's own purposes. The same holds for "good poems," "good novels," "good music," etc. in the domain of art reception. All three responses to the well-made and the good-looking include affectively positive evaluations that tend to prime motivational tendencies for prolonged and/or repeated approach/exposure.[56]

The behavioral adaptations of appreciating sexual appeal and technical mastery converge in the techniques of body ornamentation. These have a twofold potential for reward, namely, in regard to their intrinsic perfection and to their enhancement of the overall body image. Of course, techniques of body ornamentation are not least about deception. Artificial beauty is supposed to improve the natural state of the body, thus making it eligible for a degree of appreciation that would otherwise be beyond its reach. Accordingly, beautiful appearance and deceptive temptation are often closely associated.

Considering the interaction of technology/tool use and looks-based sexual preferences in (self-)decorative practices, the already mentioned archeological finding deserves all the more attention: even though human tool use has been documented for over three million years, unambiguous evidence for the technically costly production of wholly self-made three-dimensional objects of self-ornamentation is only found in the latest representative of the human line, namely the modern Homo sapiens. Moreover, this earliest evidence dates back precisely to the period (between 40,000 and 20,000 years ago) that also left behind the first non-(self-)decorative artworks. By contrast, findings of ocher and other mineral pigments in caves that were inhabited by humans suggest that forms of body painting

55 Cf. Donald, "Art and Cognitive Evolution," 6.

56 Cf. Kant, *Critique of the Power of Judgment*, 107; Hanich et al., "Why We Like to Watch Sad Films."

could have existed for over 250,000 years.[57] The extraction of pigment and the application of paint imply technologies of their own. Middle-Paleolithic ornaments found in Africa and the Near East, which are mostly about 80,000 to 120,000 years old,[58] consist of rare *natural* objects that were considered costly and/or difficult to acquire; these were perforated, often colored, and rendered applicable.[59]

As far as we know today, it was Europe in particular where the rapid artistic and technological development during the "upper Paleolithic revolution"[60] about 40,000–50,000 years before our age took place. Only since this period, there is evidence that all classic visual arts—figurative pating, sculpture, engraving—are practised in parallel and at a high technical level. All classic visual arts simultaneously, instantaneously, and at a high technical level: figurative painting, sculpture, engraving.[61] Moreover, since this time, the archeological record also shows, in great number and variety, costly ornaments whose form they completely imposed on the respective raw material (rather than taking advantage of the given shape of natural objects). This fast advancement of the arts—both of the older decorative variety and the new figurative (not immediately decorative) one—coincides with an analogous advancement in the diversity, specialization, and effectiveness of human tool cultures as documented for the period beginning ca. 50,000 years before our age.[62]

Summing up, the emergence and dissemination of these first archeologically documented arts, which (unlike the hypothetical proto-arts of singing and dancing) produced durable material objects, closely correlates with an advancement of technological capabilities that is sometimes described as a

---

57 Cf. d'Errico, "The Invisible Frontier," 197 sqq.; Barham, "Possible Early Pigment Use in South-Central Africa"; d'Errico and Soressi, "Systematic Use of Pigment by Pech-de-l'Azé Neandertals: Implications for the Origin of Behavioural Modernity"; Watts, "Ochre in the Middle Stone Age of Southern Africa"; Bar-Yosef et al., "Shells and Ochre in Middle Paleolithic Qafzeh Cave, Israel."

58 Cf. Vanhaeren et al., "Middle Paleolithic Shell Beads"; Haidle, "Wege zur Kunst," 242.

59 Cf. Bouzouggar, "82,000-Year-Old Shell Beads from North Africa"; Zilhão et al., "Symbolic Use of Marine Shells and Mineral Pigments by Iberian Neandertals"; d'Errico et al., "Nassarius Kraussianus Shell Beads from Blombos Cave"; d'Errico et al., "Additional Evidence on the Use of Personal Ornaments in the Middle Paleolithic of North Africa"; Conard, "A Critical View."

60 Cf. Mellars, *The Neandertal Legacy*; and Bar-Yosef, "On the Nature of Transitions: The Middle to Upper Paleolithic and Neolithic Revolution."

61 Cf. Floss, "Die frühesten Bildwerke der Menschheit."

62 Cf. Dunbar, *The Human Story*, 30 sqq., as well as the articles and bibliographies in the collection *Les chemin de l'art aurignacien en Europe. Das Aurignacien and die Anfänge der Kunst in Europa.*

"revolution" and sometimes as the refinement, differentiation, and systematization of possibilities that were already in principle available and occasionally used before.[63] This connection between technology and the arts does not necessarily mean that the development of new refined tools was the actual cause of the arts' emergence, or, conversely, that the need for new arts drove tool culture to new heights. Instead, Dunbar and others suggest a more complex explanatory model.[64] According to it, modern humans had already developed the current brain volume as well as the capability for syntactic and symbolic language—and thus the principle capacities for imagination, fiction, and narration—at least 150,000 years before our time, with some estimates even speaking of 500,000 years.[65] This view implies the (speculative) notion that a lengthy incubation period was necessary for these relatively new adaptations to unfold their full cumulative potential for our cognitive flexibility, our behavior as a whole, and our technical achievements.

The social brain hypothesis suggests that the new capacities were initially primarily involved in social cognition and specifically supported higher degrees of social complexity (larger groups, labor division, cooperative behavior, more precise differentiation of social orders). According to this model, the material "revolution" of tools and arts must have been preceded by a symbolic and social cognitive "revolution." Once a critical mass of social complexity, symbolic communication, and social cognitive flexibility was in place, it could promote the parallel breakthrough of new refined technologies and artistic practices that require high levels of social labor division, symbolic ways of acting, and, last but not least, technical skills. Analogous preferences for technically/aesthetically well-made objects and for good-looking bodies could then have joined forces in the production of technologically advanced (self-)decorative objects that serve both social and sexual distinctions.

These considerations suggest that the technological production of artificial decorations of the human body should be conceptualized not only in terms of direct sexual signal effects but also, and perhaps even primarily, as the implementation of systems of social symbols.[66] These systems are

---

63 The latter is Conard's reading in "Cultural Evolution in Africa and Eurasia During the Middle and Late Pleistocene."

64 Cf. Dunbar, *The Human Story*, 30 sqq.; Tattersall, "Human Origins: Out of Africa"; Zilhão et al., "Symbolic Use of Marine Shells and Mineral Pigments by Iberian Neandertals," 1027.

65 Dediu and Levinson, "On the Antiquity of Language."

66 Cf. Hovers, "An Early Case of Color Symbolism. Ochre Use by Modern Humans in Qafzeh Cave"; Zilhão et al., "Symbolic Use of Marine Shells and Mineral Pigments by Iberian Neandertals"; Tattersall, "Human Origins: Out of Africa"; Watts, "Ochre in the Middle Stone Age of Southern Africa."

sensitive both to the technological mastery of certain production techniques and materials and to the symbolic and cultural prestige of the objects thus produced. By promoting such symbol systems, the decorative arts could have provided a complex integration of technical, sexual, and social codes under the special conditions of human symbolic cognition. (The role of environmental challenges in this process is difficult to determine, all the more so as the new behavior arose not only under the stressful climate conditions of the Ice Age and in direct competition with the Neanderthals in Europe,[67] but also in numerous other habitats.)

The production of wholly artificial ornaments, as evident from the archeological record, fundamentally changed the practices of ornamentation. Whatever is painted on one's skin is inseparable from the body and the space-time of the ornamented human. Three-dimensional ornamental objects, on the other hand, can be handed down for generations or change owners and travel in time and space. They can serve as objects and/or media of diverse social transactions and accompany their owners into death as burial objects. Thus, these ornamental arts also differ markedly from the immobile paintings of the Paleolithic Era that were applied to caves and outdoor rock formations and hence are stationary. Works of music and the narrative arts are not subject to a similarly narrow confinement in space; however, they pay for their greater freedom to travel by failing to leave any archeological traces prior to the invention of writing.

It is widely assumed that Paleolithic "artists" produced their painting tools, paints, materials, and musical instruments themselves, and that they acquired and perfected their mastery in systematic exercise, probably under the guidance of more experienced painters, musicians, etc. This process enhanced cognitive, technical, and motor capacities that were decisive predictors of the quality of tool use in both craftsmen and artists. Notably, even though an advanced division of labor was practiced in both Latin and Greek antiquity, the concepts of *techné* and *ars* equally referred to the "higher" arts of painting or sculpting *and* to craftmanship in its more pragmatic meaning. This linguistic evidence also supports the notion that practical crafts and the arts that were later called the "liberal arts" exhibit substantial overlaps—or at least did so in the past.[68]

---

67 Cf. Conard and Bolus, "Radiocarbon Dating the Appearance of Modern Humans and Timing of Cultural Innovations in Europe," particularly 363–65; and Conard, "The Last Neanderthals and First Modern Humans in the Swabian Jura."

68 Analogously, Edward O. Wilson interprets the much-discussed painting practices in which chimpanzees engage in captivity when guided as merely a special manifestation of their general capacity for tool use. Cf. Wilson, "On Art," 72.

4. Symbolic Cognition/Language and the Arts

Evolutionary hypotheses regarding aesthetically preferred body ornaments and sexual courtship displays (singing, dancing, multimodal displays) are wholly restricted to *nonsymbolic signal* communication. Apparently, the capacity for symbolic cognition evolved only in humans. This capacity, in turn, includes a substantial potential for transforming older signaling practices aimed at winning over through "beauty." On top of their sexual or purely aesthetic signal value, phenomenally very similar forms of singing and dancing can acquire a wealth of additional symbolic meanings and symbol-based functions. Self-painting and ornaments can highlight (or simulate) elementary sexual signals and/or provide detailed socio-symbolic information.

Moreover, the verbal arts and the exploration of imaginary and fictional spaces and times only become possible once pure signal systems are effectively transcended. Yet again, some dimensions of pre-symbolic signaling are retained here, too;[69] for instance, acoustic/vocal signaling by largely pre-symbolic prosodic means (rhythm, melody, vocal signals of affect) is particularly strongly elaborated precisely in artistic uses of language. Thus, the *artistic elaboration* of language both pushes its symbolic capacities, and hence its differences from mere signals, to new heights and manages to preserve or reinvent the archaic affect potential of prelinguistic signaling.

The crucial question of when humans possessed linguistic capacities similar to those of current humans appears to be far from consensually solved.[70] It is widely assumed that this state dates back at least 100,000 years. However, as already indicated earlier, some experts suggest that symbolic and syntactic language could be 300,000 to 500,000 years old,[71] thus shifting its evolution well into the times of the archaic Homo sapiens. All in all, newer models suggest that the human language is evolutionarily based on (1) several very general communicative capacities (above all, theory of mind and the capacity to focus attention and intention on shared scenarios), (2) special capacities for producing sounds and gestures and (3) high capacities for vocal learning. Phenomena such as proto-linguistic utterances and holistic one-word sentences may point to evolutionary antecedents of full-blown syntactic language.[72]

---

69 Tomasello (*Origins of Human Cognition*) lays particular emphasis on this aspect.
70 For a review of the issue, cf. Fitch, *The Evolution of Language*.
71 Cf. Dediu and Levinson, "On the Antiquity of Language."
72 Cf. Wray, "Protolanguage as a Holistic System for Social Interaction."

The upcoming subsections discuss how distinctive characteristics of the fully evolved human symbolic cognition and, specifically, of human language may have shaped the human arts.

4.1 Transcending the Here and Now: Imagination and Narrativity

Gestures and sound signals of other species are very efficient in communicating current affective states, such as sexual arousal, readiness to fight or to submit to a superior animal, and for signaling the presence of danger; chimpanzees are also able to ask humans for particular things using gestures. Such gestures and sound signals attract the recipients' attention and influence their behavior. However, they are hardly ever capable to refer to *absent* objects, let alone to entities that do not exist at all in empirical terms. Words of human language, by contrast, have become independent of the actual presence (or direct imminence) of the object and/or the affective or mental state that they refer to. More than that, they generally do not refer to any object or external event but to a *signatum* (i.e., a meaning which is itself a product of the human mind rather than an ordinary object). This applies to all types of human signals as distinguished by Peirce (i.e., icons, indexes, and symbols).[73] Therefore, unless the context clearly indicates otherwise, the following considerations regarding human *symbolic* representation and cognition do not aim at distinctions within different types of human sign use, but are exclusively focused on distinguishing human sign use from animal signaling.

Studies on features of human cognition, rites, and gestures suggest possible preverbal predecessors of verbal language.[74] According to this view, before the evolution of a full-blown syntactic language, reports of successes and failures in hunting or fights with neighboring groups must have been primarily gesture-based, probably supported by the vocal expressions already available. Even at this early stage, such prelinguistic acts of representation would have involved an aspect of human symbolism that sets it apart from animal signals: namely, they refer to things and events of the *past*. Repeated uses of such gestural reports might have led to stabilizing specific abstract "concepts" associated with them and storing them in

73 Peirce, *Semiotics and Significs*.

74 Cf. Molino, "Toward an Evolutionary Theory of Music and Language," 174 sqq.; Richman, "How Music Fixed 'Nonsense' into Significant Formulas: On Rhythm, Repetition, and Meaning"; Turner, *The Literary Mind*, esp. 140–68.

memory. Such decoupling from an object's or event's immediate presence opens up a sphere of mental images.

Variants of such multimodal representations and of their ritual-like performance can be observed in chimpanzees. For instance, in the state of affective agitation before an attack, whole groups engage in particular combinations of sound and movement patterns. Human rituals both directly continue such collective practices of emotional self-regulation (e.g., in songs and the motor synchronization before or during military campaigns) and also increasingly use them for symbolic purposes, such as enacting mythological or religious narratives.

As syntactic human languages of the contemporary type evolved and came to dominate social communication, the preserved older forms of primarily gestural symbolization and representation (partly accompanied by nonverbal vocalizations) probably changed in status and function. Henceforth, they were no longer the key players of social communication, but accounted only for an increasingly limited fraction of it. Their aesthetic elaboration and their theater-like quality (a pantomime without words, yet in some cases accompanied by nonverbal sounds) continued and continues to play a role aside from and in addition to verbal language. Importantly, these older forms of communication were preserved in the long term, suggesting that they serve nonredundant representational and communicative needs and cannot be fully replaced by verbal language. Moreover, prelinguistic symbolism is re-experienced by every child. It also remains embedded in verbal language as emotional prosody, rhythm, and rhetorical figurativeness; as poetry, it connects even the most highly developed verbal language back to a sensitivity for sound qualities and sound-iconic meanings.

Human verbal and nonverbal symbol use radically transcends the limitation of only representing present experiences. It can address the whole range of past, present, and future events, and can even refer to things that do not exist in reality, and perhaps never will. Breaking out of the here and now, entering the dimensions of "what is no more," "what will be," "what might be," and "what never was or will be," profoundly expands our cognitive and imaginative reach.[75] The schema-building power of imagination is required already for the mental representation of real events, and all the more for remembering or conjuring up images of past and future. In the final analysis, it is the symbol-based possibility of addressing the past and future that *creates* time—in the sense of its conscious mental representation and experience—in the first place.

75 Cf. Boyd, *On the Origin of Stories.*

Darwin regards "*Imagination*" as "one of the highest prerogatives of man."[76] Here, he follows poetry-related theories of imagination developed by the German Romantics who, in their turn, followed Kant:

> A poet, as Jean Paul Richter remarks, "who must reflect whether he shall make a character say yes or no—to the devil with him; he is only a stupid corpse." Dreaming gives us the best notion of this power; as Jean Paul again says, "The dream is an involuntary art of poetry." The value of the products of our imagination depends of course on the number, accuracy, and clearness of our impressions, on our judgment and taste in selecting or rejecting the involuntary combinations, and to a certain extent on our power of voluntarily combining them. As dogs, cats, horses, and probably all the higher animals, even birds have vivid dreams, and this is shewn by their movements and the sounds uttered, we must admit that they possess some power of imagination.[77]

The difference between the human power of imagination and that of the animals referred to by Darwin is probably based on the fact that our expressive and representational capacities are not limited to direct bodily expressions (through voice, gestures, and other movements), but are greatly expanded by the use of external instruments and media, such as paint, utensils for painting and carving, applying artificial ornamental objects to the body, writing and other notation systems. These external media enormously increased the potential, both voluntary and involuntary, for *combining real, given percepts with constructs of possible events and entities.*

Narratives are the canonic form of imaginative memory, experience, and fiction. Even in their prelinguistic, theatrical form, narratives imply the creation of cognitive categories and complex relations between these categories—relations that presuppose or at least lay the path for symbolic representations. The hypothesis of an essential narrativity of the human mind finds much support today,[78] not least in developmental psychology. Across different cultures, children learn to understand and produce narratives in very similar steps and time windows, once they have learned verbal

76 Darwin, *Descent of Man*, I 45.

77 Darwin, *Descent of Man*, II 44 f.

78 Cf. Porter, "The Evolutionary Origins of the Storied Mind"; Turner, *The Literary Mind*; Turner, *From Ritual to Theatre*, 68–77; Carroll, *Literary Darwinism*; Gottschall and Wilson, *The Literary Animal*; Sperber and Hirschfeld, "The Cognitive Foundations of Cultural Stability and Diversity"; Gottschall, *The Storytelling Animal.*

language and thus gained access to symbolic thought. Elementary human cognitive patterns likewise appear to be narrative in character.

Artistic elaborations of narrative patterns used in real life contexts are thus potentially of great importance for the ontogenetic development of our narrative mind. A widespread consensus stipulates that a narrative contains (at least) one protagonist, a setting, a developing plot, in which the protagonist encounters conflicts and complications on the way to his or her goal, and a denouement (an ending).[79] Narratives historically "selected" as myths, fairy tales, etc. differ from narratives about the many minor everyday problems and conflicts by their tendency to increase the scope and affective salience of the goals, conflicts, and possible rewards (corresponding to Darwin's principle of aesthetic "exaggeration"). This renders these narratives particularly memorable. The narrative plots of tragedies condense the emotional conflicts defining entire lives of outstanding protagonists into a performance time of just two to four hours. Readers of novels likewise often go through entire life trajectories, albeit mostly in reference to the lives of more usual protagonists, within three to fifty or so hours of reading.

Compared to everyday plots, "artistic" narratives not only increase the content-based emotional salience, but also rely more strongly on stylistic means of message intensification and winning over the audience, as conceptualized by Greek and Latin rhetoric. Narrative art thus makes aesthetically ambitious use of the general human capacities for narration. Therefore, it does not seem necessary to stipulate special evolved "modules" for narrative art forms. Rather, we can share Ellen Dissanayake's view of artistic narration as a "making special"—a supernormal elaboration—of a general cognitive adaptation.[80]

So far, a great part of the research on literature that is informed by evolutionary reasoning has been limited to the content of epics, novels, dramas, and poems.[81] Unquestionably, all forms of literature primarily deal with social conflicts, life trajectories, and desires of human protagonists. The topic of seeking and obtaining a sexual partner (mostly for marriage)—and the substantial efforts and social conflicts involved in this quest—are of preeminent importance in this context.[82] Thus, narratives tend to have a strong focus on the very issue that evolutionary psychology has identified as particularly "hot" and salient for all species, namely, reproductive success.

79 Cf. Sugiyama, "Reverse-Engineering Narrative," 180.

80 Dissanayake, *What Is Art For?*

81 For a more extended discussion of this issue, see Davies, *The Artful Species*, 163–76.

82 Cf. Dutton, *The Art Instinct*, 132.

According to such purely content-based criteria, hospital romances and similar products should have the status of artworks per excellence as they deal with "sexual choice" most directly and often exclusively, and as their readers are plentiful.[83] For classical scholars of literature, many literary studies informed by evolutionary psychology ("evocrit") are therefore too content-focused and aesthetically undiscriminating. However, such shortcomings can be easily avoided in light of Darwin's theory of the arts. Darwin never expressed the expectation that scenarios of sexual selection should be always literally represented by the arts. Instead, he sees the aesthetic and emotional power of the arts in the way their particular *formal characteristics*—which he consistently refers to in terms of musical tones, rhythms, and cadences[84]—mobilize affective energies that associatively remind people of an archaic sexual heritage, without the artists and audiences necessarily being consciously aware of this association.

In general, individual artistic narratives share the distinctive characteristics of content intensification (supernormal manifestations of conflicts, fights, suffering, and rewards) and heightened musicality with great collective narratives. At the same time, they can produce many new aesthetic effects by virtue of the high degrees of artistic freedom regarding how the underlying plot is transformed into the actual narrative. This can, for instance, imply that attention-heightening effects are achieved by partially violating habitual expectations of how a narrative unfolds.

In everyday contexts, the narrativity of our mind is not necessarily fictional and imaginary in nature. The cognitive and even affective involvement with purely fictional beings and imaginary values is, according to current knowledge, a distinctive human capacity dependent on human symbolic cognition. Artistic and religious imagination appears to be highly dependent on this special imaginative ability.[85] Among the very oldest figurative paintings preserved (40,000 to 15,000 years before our time), we find flights of fancy such as hybrids between animals and humans (for instance, lion-men in upright walk/stand position) as well as between different animal species. According to the theory advocated here, such fictional, nonempirical beings could only be created by the human mind after the evolution of the capacities for symbolic understanding and communication, which also underlie language.[86]

---

83 Cf. Whissel, "Mate Selection in Popular Women's Fiction."

84 Darwin, *Descent of Man*, II 336–37.

85 Cf. Darwin, *Descent of Man*, I 65.

86 Cf. Boyd, "The Evolution of Stories."

The conceptual blending theory developed by Mark Turner and other cognitive linguists provides a cognitive linguistic perspective on these phenomena.[87] One of the central achievements of human language is the capacity to communicate a principally infinite array of semantic nuances by means of a finite lexical reservoir. This is possible due not only to ever new syntactic combinations but also to the constant creation of new terms and concepts via the metaphorical use of existing words. From a cognitive perspective, figurative speech is a superimposition or blend of a source domain and a target domain. These terms stress that metaphor is not merely a matter of replacing one semantic meaning (the literal one) by another (the figurative one), but also of a transfer between different senses and cognitive domains. Cognitive blends such as "the foot of a mountain" or "the journey of life" are elementary and extremely frequent language-based methods of creating meaning. Ancient sculptures blending human and animal parts can be understood as visual analogues of such linguistic blends. Featuring talking foxes and other curious creatures, the genre of the fable makes topical use of such hybrid beings.

However, there is also an important difference between animal-human hybrids and analogous phenomena on the one hand, and metaphorical linguistic blends on the other. Whereas the former violate natural and/or culturally accepted ontologies, the latter do not. Equating "life" with "journey" requires a nonliteral processing of the word "journey"; thereby, the mind overcomes the perceived mismatch of the literal word meaning and the broader sentence context. A human-animal hybrid, on the other hand, retains a stronger unresolved alterity, even if interpreted as some sort of metaphor. It is precisely this ontological deviance that is used across cultures as a marker of mythical and religious beings and values. Neuroscientific research suggests that counterintuitive blends that infringe on accepted ontologies and add paranormal levels (the supernatural, the divine) to our "normal" world are processed with particular attention and enjoy privileged access to as well as retrieval from memory. They are, then, predestined for better survival as cultural memes than more "normal" blends. This is why all religions and mythologies make systematic use of superimpositions of this ontology-violating type.[88]

---

87 Cf. Turner, "The Cognitive Study of Art, Language, and Literature"; and "The Art of Compression."

88 Cf. Barrett and Nyhof, "Spreading Non-Natural Concepts"; Boyer, "Religious Thought and Behavior as By-Products of Brain Function"; Boyer and Ramble, "Cognitive Templates for Religious Concepts."

Basic linguistic processes thus support the emergence of para-natural beings in human culture. By doing so, they promote the creation of ideologies and systems of thought that feature imaginary beings as if this were the most natural thing in the world. From an evolutionary perspective, it is hardly an exaggeration to say that the human being is the specialist for the absent, the imaginary, for things that transcend experience and the present tense—in short, for the wide field of sign-based "productive imagination." In Kant's words, this imagination is "very powerful . . . in creating another kind of nature" and systematically "strives" for "something beyond the boundaries of experience."[89] The pathos of this statement suggests that the disposition toward transcending ordinary reality seems to be part and parcel of the very definition of humanity. What, then, are the evolutionary advantages of our enormous susceptibility to and disposition for imagining fictional entities? Why was it precisely "the gullible ape,"[90] probably the only species with an out-of-control penchant for creating imaginary worlds, that managed to achieve such overwhelming superiority over all other creatures on Earth?

The classic answer given by poetics and theories of fiction is: this capacity allows us to mentally explore a great range of possibilities (including highly improbable ones). If addressing the possible and even the impossible is a distinctive option of our symbol use, then artistic-imaginative practices make especially intensive uses of this option. They make it possible to imaginarily act out optional scenarios in a great variety of ways. Not every single instance of this behavior is necessarily advantageous; on the whole, however, individuals and whole cultures apparently need a reservoir of symbolic variations and orientations in order to be able to respond to new challenges in a flexible fashion.

Narratives and other works of art that provide such means can hardly be conceived of as resulting from or directing sexual choice. If narrative brilliance were exclusively driven by sexual selection, it would entirely fulfill its function by impressing the other sex. There would be no reason why it should enable the addressee to consider different courses of future action in imagination or to build imaginary social mythologies. According to Darwin, the peacock fan or proto-music have no further use apart from their purely sexual signal character. Therefore, the general human

89 Kant, *Critique of the Power of Judgment*, 192. Cf. Harris, "The Work of the Imagination."

90 Cf. Power, "'Beauty Magic': The Origins of Art," 95. The memorable image of "the gullible ape" was not coined by Camilla Power herself; unfortunately, I did not succeed in finding the source. Cf. also Wilson's reflections on "indoctrinability" as a human behavioral characteristic in *Sociobiology*, 562.

capacities of symbolic cognition (including fiction, simulation, etc.) are largely regarded as a result of natural rather than sexual selection.

However, the relative disempowerment and the transgression of the one and single reality is only one side of the functional theory of imaginary worlds; it is the side habitually known as the modern ideology of the arts. Journeys into other, paranormal worlds can also serve the opposite goal, namely to consolidate the "normal" world by the very act of taking an imaginative detour.[91] Ethnological theories tend to place a primary focus on this art-supported construction of a collective imaginary that stabilizes norms, values, and narratives, thus promoting social cohesion and cooperation. (This view has already been discussed in Chapter 2.)

4.2 Tolerance for and Competence in Ambiguities and Indeterminacies

The great majority of signals produced by other species is unambiguous. Most appear to be associated with a clear and stable meaning. In regard to more flexible vocalizations, additional context-based variables tend to ensure correct decoding. In any case, animal signals are hardly mysterious messages that require time-consuming interpretive efforts, even when they work with graded rather than categorical or with apparently ambiguous signals.[92] Human language, on the other hand, is less connected to real, present objects and situations, more self-referential, and far more imagination-friendly—and thus also more in need of interpretive efforts. For the same reasons, it is also particularly capable of ambiguity/plurivalence and of practices of deception.

From the perspective of pragmatic communication, this can be seen as a negative side effect of the extreme flexibility with which we use words, the high degree of freedom we have to (re-)combine them, and their equal capacity to refer to non-real and real entities. From the perspective of mythologies, religions, and also politics, on the other hand, the capacity of language to evoke and support complex ambiguities—including "dark and mysterious" levels of meaning, variable reinterpretations, and multifold illusionary effects—is highly useful. Without these cardinal capabilities

---

91 Another strongly conservative interpretation of our acceptance of religious narratives, political ideologies, and fictions of all kinds states that this acceptance is simply a socially adaptive strategy of mental conformism. Within groups and hierarchies, such behavior is stipulated to create more advantages than disadvantages (Wilson, *Sociobiology*, 562).

92 Cf. Wilson, *Sociobiology*, 176–93.

of our symbolic thinking, a wide field of cultural phenomena would be literally unimaginable.

The individual arts include special predispositions for pushing the general potential of symbolic representations for ambiguity and allusive resonance to higher levels. Some of them (according to Darwin's theory, the ancient sexually selected arts of song and dance in particular) have a strong pre-symbolic heritage, which is by definition in tension with symbolic representation. They aim precisely for the opposite of symbolism, namely, for the sensual and affective elaboration of the sign's material body rather than its reduction to a wholly arbitrary and phenomenally irrelevant vehicle of some abstract meaning. As Roman Jakobson famously puts it, the sign's self-referentiality (its "poetic function") generally tends to render messages more ambiguous.[93]

This consequence becomes all the clearer if we consider the difference between calls and songs. As discussed above, songs are rich in variation and often require long periods of practice; they usually take up more time than calls, are characterized by melodic complexity and largely resist decoding and semantic interpretation. Especially this latter characteristic must almost inevitably lead to interpretive and associative efforts on the part of a species with high symbolic capacities and a strong proclivity to search for "meaning." With regard to art, such efforts can typically never exhaustively integrate the complexities of the concrete aesthetic material into a meaning abstracted from it. Thus, artistic practices significantly increase the capacities of symbolic communication for activating a broad range of associative resonances.[94] This principal semiotic quality of the arts fits in well with Darwin's hypothesis of the emotional resonance of proto-sexual yet only vaguely decoded feelings in music and language.

The function of these cognitive complexities of representation combined with heightened sensual and affective salience can be elucidated by reference to basic characteristics of human social life. We humans are evolutionary generalists par excellence: thanks to elaborate social and cognitive strategies, we achieve highly cooperative performances in many areas. We have proven to be flexible and competitive enough to survive in almost any environment, be it artificially created or naturally found on Earth. However, this flexibility comes at a price: we lack the guidance of the highly

---

93 Jakobson, "Linguistics and Poetics," 370–71. For an empirical proof of this assumption, see Wallot and Menninghaus, "Ambiguity Eeffects of Rhyme and Meter;" see also the numerous reflections on ambiguity and the rich connections between sounds and meanings in Schrott and Jacobs, *Gehirn und Gedicht*.

94 Cf. Zeki, "The Neurology of Ambiguity."

specialized behavioral adaptations that create a narrow and unambiguous correspondence between an organism and its habitat. Being linguistically and symbolically gifted, we can make up for this potential downside of our flexibility by creating systems of symbolic orientation (myths, religions, ideologies); these offer synthetic "meanings" which make certain actions and interpretations more likely than others. This way of coping with the contingencies of human existence requires high-order communicative capacities and reflective loops that go far beyond the encoding and decoding of unambiguous signals.

Darwin defined conscience as an "inward monitor"[95] that registers and processes the conflicts between social and egoistic impulses. What makes the human "moral sense" especially acute is the fact that the expectations of communities—and our conflicts with them—are elaborately negotiated on the stage of our consciousness. According to E. O. Wilson, conflicts of ambivalence are more frequent, basic, and inevitable in human social life than in any other species.[96] As conscious individuals granted great behavioral freedom, we act with more conscious egotism than individuals of other species. On the other hand, this tendency is continuously balanced out by high, complex, and very flexible (compared to other animals) demands for social acceptance. These, too, determine our behavior not only on a subconscious level but are also reflectively represented in our consciousness. Thus, we become subject to external and internal conflicts, in which we are capable of taking either side. This is why we need high levels of tolerance for ambiguity and unresolved conflicts—but also forms of thought that can occasionally provide these cognitive and affective challenges with a self-rewarding turn or even a temporary "solution."

Artistic narratives, images, and sounds might be this much-needed resource: they combine a high complexity and potential for not fully determinate meanings—what Baumgarten calls the *ubertas* or *copia*, the richness of resonances by means of which *aesthetic cognition* departs from the precision and distinctness of theoretical cognition[97]—with a high affective salience and inherent processing reward. Culturally successful narratives, especially myths, often admit two or more interpretations. Thus, they flexibly serve needs for reflection and orientation which can differ strongly both diachronically and synchronically (the latter dependent on the respective social status). Kant stressed the principal impossibility of conceptualizing

95 Darwin, *Descent of Man*, I 73.

96 Wilson, *Sociobiology*, 129, 563.

97 Cf. Baumgarten, Ästhetik and *Meditationes Philosophicae de Nonnullis ad Poema Pertinentibus.*

"aesthetic ideas" in a fully determinate fashion and likewise correlated their cognitive processing with indeterminacy competence as well as with complex, potentially indefinite interpretive efforts.[98]

4.3 The Risks and Potentials of Deception and Self-Delusion

The concept of deception—and the related ideas of "illusion" and "as if" in art—is a basic concept of rhetoric, poetics, and aesthetics. Purely visual and acoustic illusions can well be devoid of genuine symbolic meaning. The main domain of deception, however, involves symbolic cognition.[99] The pre-Socratic philosopher Gorgias has written one of the earliest apologias of deception. Plutarch reports:

> Tragedy blossomed forth and won great acclaim, becoming a wondrous entertainment for the ears and eyes of the men of that age, and, by the mythological character of its plots, and the vicissitudes which its characters undergo, it effected a deception wherein, as Gorgias records, "he who deceives is more honest than he who does not deceive, and he who is deceived is wiser than he who is not deceived." For he who deceives is more honest, because he has done what he has promised to do; and he who is deceived is wiser, because the mind which is not insensible to fine perceptions is easily enthralled by the delights of language.[100]

Immersing oneself in the illusion and deception of a tragic performance—instead of disregarding it as unserious and unreal—thus creates the affective advantage of experiencing word-induced pleasure and also the cognitive advantage of "being wiser." Without affective involvement, there would barely be any illusion of the imaginary reality represented, and without accessibility to some cognitive understanding, this illusion would be less pleasurable. Thus, the dual innervation of cognitive and affective processes supports aesthetic pleasure and can also enrich cognitive/affective strategies of action and interpretation.

According to evolutionary theory, deception strategies are important adaptive behaviors of countless creatures. Typically, such strategies are engaged in a continuous arms race with the equally adaptive counterstrategies of deception detection. As regards human communication, the form of drama, both in its tragic and comic varieties, offers particularly rich opportunities for

98 Kant, *Critique of the Power of Judgment*, 192–95.
99 Cf. Joyce, *The Esthetic Animal*, 35–39.
100 Plutarch, *De Gloria Atheniensum* 5.

learning more about deception strategies in social interaction and for refining both one's deception and one's deception detection skills.[101]

However, this exercise in social detection skills certainly does not exhaust the potential of the arts. The great need and capacity for individual and collective *self*-delusion appears to be a specific human quality. Viewed soberly, both social ideologies and individual self-concepts often are systems of (at best) partial understanding, or even of (willingly or unwillingly) deceiving others and oneself.[102] Self-delusion can be functionally advantageous in as far as it renders the deception of others more credible and thus more successful,[103] but also as an antidote to "depressive realism."[104]

At the same time, the evolutionary selection of behavioral patterns does by no means imply that the respective agents are fully aware of the motives underlying their actions.[105] This rule is not invalidated by the fact that we humans often superimpose conscious motivations on our evolved behavioral mechanisms. According to Richard Alexander, the conscious awareness of one's own actions immediately leads to a network of (self-) deceptions and resistances against a soberer insight on the one hand, and an almost impenetrably unconscious dimension of our actions on the other.[106] This diagnosis of principally incongruous human behavioral systems bears an astonishing resemblance to Freud's hypotheses regarding the systems of the conscious and subconscious. Both Alexander and Freud diagnose a great need for strategies of self-delusion as well as complex functional mechanisms of misinterpreting others. Against this background, aesthetic practices of suspending one's disbelief for the illusions of art *also* offer opportunities for largely turning off self-criticism and for basking in pleasurable illusions. Drug-induced and other trances are additional transculturally widespread practices often serving similar functions.

From an evolutionary perspective, such mechanisms of self-delusion can well be functional adaptations if their overall net effect is more conducive than detrimental to the well-being of individuals and/or social groups. After all, nothing is worse than to be robbed of all illusions, and it is precisely literature that shows this very clearly. For humans, who systematically

101 Cf. Hansen, "A Prehistory of Theatre," 359. On narrative, cf. Sugiyama, "On the Origins of Narrative."

102 Cf. Alexander, "The Search for a General Theory of Behavior," particularly 96 sqq.; Wilson, *Sociobiology*, 119 and 553; Cook, "Edward O. Wilson on Art," 102 and 114.

103 Cf. Trivers, "Deceit and Self-Deception."

104 Alloy and Abramson, "Depressive Realism: Four Theoretical Perspectives."

105 Alexander, "The Search for a General Theory of Behavior," 96 sqq.

106 Cf. Alexander's numerous reflections on this problem in *Darwinism and Human Affairs*.

interpret themselves in terms of imaginary values, worlds, and beings, illusions are not only a cognitive problem but also a valuable resource. The arts feed and support this resource, even if also at times questioning it.

5. The Four Coopted Adaptations in Interaction

Typologically, the hypothesis on art that has been sketched in this chapter corresponds to a general hypothesis about the structure of the human mind. According to it, our peculiar mental creativity and flexibility is based to a large degree on the cross-modal use we make of our cognitive, emotional, and behavioral capacities and dispositions.[107] Many evolutionary processes take the path of reusing or coopting already existing adaptations. This corresponds to the principles of parsimony and gradual variation.

The four adaptations that have been proposed as preconditions and forerunners of the human arts are of different ages and appear to have been largely or completely independent from one another for a very long period. Only the presumably youngest of these adaptations—human symbolic language, the main motor of our cognitive fluidity—appears to have created the conditions required to transform, or to reconfigure, the three older adaptations—the aesthetic (and partly sexualized) sensibility for differences in the beauty of looks, the pleasures of play, and tool use—into what we today call "the arts," or even "art." Thus, my thesis is: *The human arts arose as new variants of human behavior, as three ancient and largely independent adaptions—the sensitivity for differences in visual and auditory beauty, play behavior, and tool use—joined forces with, and were transformed by, the human capacities for symbolic cognition and language.*

This hypothesis implies that important components of artistic behavior precede our symbolic capacities and thus are *not* a function of these capacities. This applies specifically to the sensitivity for purely material and perceptual properties independent of any symbolic meaning, and hence to the first component of the cooptation model: the "sense of beauty" in its primary evolutionary meaning. The fundamental role of this component also explains why language and symbolic capacities—even though they contributed decisively to the emergence of the arts during the Upper Paleolithic Era—often reach their limits when trying to account for aesthetic experiences. As Baumgarten and Kant were the

107 Cf. the reflections on our higher cognitive capacities in Fodor, *Modularity of Mind*; Donald, "Art and Cognitive Evolution," 17; and Mithen, *The Prehistory of the Mind*. See also Neumann, *Funktionshistorische Anthropologie der ästhetischen Produktivität*, particularly 117.

first to emphasize, the arts confront symbolic language with sensuous and material dimensions that partly resist any exhaustive translation into the language of concepts. It is this partial divergence of symbol-supported theoretical cognition and aesthetic perception/evaluation that motivated eighteenth-century philosophers to found a new discipline called "aesthetics." Rather than destroying this much-discussed "autonomy" of aesthetic appreciation, evolutionary aesthetics has unique potential to give it new theoretical substance.

Notably, none of the components of the cooptation hypothesis—"sensitivity for (sexual) beauty," "play behavior," "tool use," and "symbolic cognition"—are exclusively drawn on in the production and reception of the arts. Moreover, the new use of these adaptations in the field of the arts does not (necessarily) imply that this new function replaces the respective old ones. Rather, the old functions usually persist independently of the arts.[108] At the same time, the great importance of the adaptations which were recruited for artistic behavior and aesthetic enjoyment does not at all exclude the possibility that particular arts might exhibit some highly specific characteristics that *cannot* be understood by reference to the hybrid cooptation of a given pool of domain-general capacities. In the following, this possibility shall be briefly discussed in regard to different arts.

Darwin's construct of our distant ancestors' sexually courting singing fits in best with the concept of a domain-specific adaptation as defined by the narrow standards of evolutionary biology. It can remain open here if the arts of singing fully owe their evolution to a sexual function, or if they are partly based on other mechanisms that had originally evolved for other functions. If such mechanisms were subject to further development and refinement for purposes of sexual singing, then, at least in principle, both cases would qualify as specialized adaptations. Based on these premises and following Darwin, one could well consider the human evaluative sensitivity for "high musical capacities"[109] as having originally evolved as a second specialized mechanism of sexual courtship alongside the sensitivity for advantages in looks. This assumption is model-conform, free of internal contradictions and, so far, empirically unrefuted. It is also by no means irreconcilable with the cooptation hypothesis advanced here. However, this hypothesis has not been proven true, either.

Compared to music, visual self-ornamentation offers far fewer indications suggesting an origin as a modular, task-specific adaptation. According to Darwin, the decorative arts originally supported sexual and

108 Cf. Gould and Vrba, "Exaptation."

109 Darwin, *Descent of Man*, II 335.

social self-presentation as well as the display of natural charms of bodily looks. Thus, the decorative arts recruit our visual system and our tool use capacities for functions that are at least partly derived from the sexual evaluation of natural, body-based aesthetic virtues. If continued practices of the decorative arts did not lead to *specialized and genetically fixed optimizations* of one of these characteristics (dispositions and preferences of the visual system, technological capacities)—and, so far, there seems to be no evidence for such a development—these arts can readily be conceptualized as a culturally developed (re)use of existing adaptations.

Regarding the verbal arts, it is also an open question whether or notthey involve some characteristics of a domain-specific adaptation. Even if our *narrative capacities* have a special genetic basis (as the ontogenetic regularity of its acquisition suggests), there appears to be no need to assume special additional adaptations for *artistic* narratives.

The case of *rhetorical capacities*, which Darwin conceived as an affective music-related elaboration of language itself, is less clear. Does oratorical and poetic talent of this type have specially evolved genetic foundations that limit the opportunities for refining such capacities by learning? No clear answer to this question is in sight. On the whole, rhetorical capacities have barely been studied either in regard to their ontogenetic development or to possible phylogenetic foundations that are distinct from those underlying the production of language overall. Apart from the general difficulty of identifying genetic correlates for highly complex capacities, an institutional reason is also at work here: rhetoric simply has no firm place anymore within the grid of contemporary disciplines. Though rhetorical capacities play an enormous role in social communication, recent psychology and linguistics barely investigate them.

Forms of song, on the other hand—not intra-linguistic analogies to musical prosody, but direct combinations of language and music—are hardly in need of any special additional adaptation beyond the learnable interaction of music and language. Analogously, it appears to be highly unlikely that the words and gestures of the theatrical arts rely on special innate adaptations beyond those of our body language reservoir, our linguistic competence and our ability to express emotions vocally (prosodically).

No matter what the exact answers to such pending questions are going to be, the cooptation model of the human arts advocated here stresses the importance of capacities and dispositions that did not modularly evolve for the sake of the arts but were coopted, (re)configured and (re-)deployed in the service of art production and reception. The model expands Darwin's criterion of learnable variance into the following hypothesis: it is not (or not only) a single specialized capacity that can be driven to artistic heights; rather, producing and experiencing the arts represent outstanding achievements resulting from the joint employment and novel uses of previously separate adaptations across multiple domains. This model accords with the widely shared notion that the peculiar creativity and flexibility of the human mind specifically manifests itself in such cross-modular uses of its special evolved capabilities. It provides a new evolutionary variant of Kant's famous hypothesis of a "free (inter)play of our capacities" in "aesthetic pleasure."[110]

110 Kant, *Critique of the Power of Judgment*, 102–3.

Bibliography

Abbott, H. Porter. "The Evolutionary Origins of the Storied Mind: Modeling the Prehistory of Narrative Consciousness and Its Discontents." *Narrative* 8, no. 3 (2000): 247–56. Accessed August 7, 2017. doi:10.2307/20107217.

Abel, Julia, and Ralf Stürmer. "Aristoteles im Test. Psychophysiologische Untersuchungen zur Wirkung von Tragödien." In *Heuristiken der Literaturwissenschaft. Disziplinexterne Perspektiven auf Literatur*, edited by Uta Klein, Katja Mellmann, and Stefanie Metzger, 13–34. Paderborn: Mentis, 2006.

Abelson, Robert P. "Psychological Status of the Script Concept." *American Psychologist* 36, no. 7 (1981): 715–29. Accessed July 24, 2017. doi:10.1037/0003-066X.36.7.715.

Aiello, Leslie C. "Terrestriality, Bipedalism and the Origin of Language." In *Evolution of Social Behaviour Patterns in Primates and Man*, edited by Walter G. Runciman, John M. Smith, Robin I. M. Dunbar, Society Royal and Academy British, 269–90. Oxford: Oxford University Press for the British Academy, 1996.

Aiken, Nancy E. *The Biological Origins of Art*. Westport, CT: Praeger, 1998.

Aitken, Paul P. "Judgments of Pleasingness and Interestingness as Function of Visual Complexity." *Journal of Experimental Psychology* 103, no. 2 (1974): 240–44. Accessed August 4, 2017. doi:10.1037/h0036787.

Alexander, Richard D. "Natural Selection and Specialized Chorusing Behavior in Acoustical Insects." In *Insects, Science and Society*, edited by David Pimentai, 35–77. New York: Academic Press, 1975.

Alexander, Richard D. "The Search for a General Theory of Behaviour." *Behavioral Science* 20, no. 2 (1975): 77–100. Accessed July 24, 2017. doi:10.1002/bs.3830200202.

Alexander, Richard D. *Darwinism and Human Affairs*. Seattle: University of Washington Press, 1979.

Alloy, Lauren. B., and Abramson Lyn. "Depressive Realism: Four Theoretical Perspectives." In *Cognitive Processes in Depression*, edited by Lauren B. Alloy, 223–65. New York: Guilford Press, 1988.

Ambers, Janet C. "Raman Analysis of Pigments from the Egyptian Old Kingdom." *Journal of Raman Spectroscopy* 35 (2004): 768–73.

Anati, Emmanuel. *Les Origines de l'Art et la Formation de l'Esprit Humain*. Paris: Albin Michel, 1989.

Anati, Emmanuel. *World Rock Art. The Primordial Language*. Brescia: Centro Communo di Studi Preistorici, 1993.

Anshel, Anat, and Kipper David A. "The Influence of Group Singing on Trust and Cooperation." *Journal of Music Therapy* 15 (1988): 145–55. Accessed August 8, 2017. doi:10.1093/jmt/25.3.145.

Apter, Michael J. "Reversal Theory, Cognitive Synergy and the Arts." In *Advances in Psychology: Cognitive Processes in the Perception of Art*, edited by W. Ray Crozier and Antony J. Chapman, 411–26. North-Holland, Amsterdam: Elsevier Science Publishers B. V., 1984.

Arcadi, Adam C., Robert Daniel, and Christophe Boesch. "Buttress Drumming by Wild Chimpanzees: Temporal Patterning, Phrase Integration into Loud Calls, and Preliminary Evidence for Individual Distinctiveness." *Primates* 39 (1998): 505–18.

Arcadi, Adam C., Robert Daniel, and Mugurusi Francis. "A Comparison of Buttress Drumming by Male Chimpanzees from Two Populations." *Primates* 45, no. 2 (2004): 135–39. Accessed August 8, 2017. doi:10.1007/s10329-003-0070-8.

Augustin, M. Dorothee, Johan Wagemans, and Claus-Christian Carbon. "All is Beautiful? Generality vs. Specificity of Word Usage in Visual Aesthetics." *Acta Psychologica* 139, no. 1 (2012): 187–201. Accessed October 9, 2017. doi:10.1016/j.actpsy.2011.10.004.

Augustin, M. Dorothee, Claus-Christian Carbon, and Johan Wagemans. "Artful Terms: A Study on Aesthetic Word Usage for Visual Art versus Film and Music." *I-Perception* 3, no. 5 (2012): 319–37. Accessed August 10, 2017. doi:10.1068/i0511aap.

Aunger, Robert, ed. *Darwinizing Culture: The Status of Memetics as a Science*. Oxford: Oxford University Press, 2000.

Aunger, Robert. "Conclusion." In *Darwinizing Culture: The Status of Memetics as a Science*, edited by Aunger, Robert, 205–32. Oxford: Oxford University Press, 2000.

Bakhtin, Michail M. *Rabelais and His World*. Translated by Hélène Iswolsky. Bloomington: Indiana University Press, 1984.

Barham, Lawrence S. "Possible Early Pigment Use in South Central Africa." *Current Anthropology* 39 (1998): 703–20.

Barrett, Justin L., and Melanie A. Nyhof. "Spreading Non-Natural Concepts: The Role of Intuitive Conceptual Structures in Memory and Transmission of Cultural Materials." *Journal of Cognition and Culture* 1, no. 1 (2001): 69–100. Accessed August 4, 2017. doi:10.1163/156853701300063589.

Barkow, Jerome H., Leda Cosmides, and John Tooby, eds. *The Adapted Mind: Evolutionary Psychology and the Generation of Culture*. Oxford: Oxford University Press, 1992.

Barthes, Roland. "Leaving the Movie Theatre." Translated by Richard Howard. In *The Rustle of Language*, 345–50. Berkeley: University of California Press, 1986.

Barthes, Roland. *The Neutral: Lecture Course at the College de France (1977–1978)*. Translated by Rosalind Krauss and Denis Hollier. New York: Columbia University Press, 2007.

Bar-Yosef, Ofer G., and Bernard Vandermeersch. "Modern Humans in the Levant." *Scientific American* 268, no. 4 (1993): 94–100.

Bar-Yosef, Ofer G. "On the Nature of Transitions: The Middle to Upper Paleolithic and Neolithic Revolution." *Cambridge Archaeological Journal* 8, no. 2 (2008): 141–63. Accessed July 26, 2017. doi:10.1017/S0959774300001815.

Bar-Yosef Mayer, Daniella E. Bernard Vandermeersch, and Ofer Bar-Yosef. "Shells and Ochre in Middle Paleolithic Qafzeh Cave, Israel: Indications for Modern Behavior." *Journal of Human Evolution* 56, no. 3 (2009): 307–14. Accessed August 10, 2017. doi:10.1016/j.jhevol.2008.10.005.

Baumgarten, Alexander G. *Meditationes Philosophicae de Nonnullis ad Poema Pertinentibus, Philosophische Betrachtungen über einige Bedingungen des Gedichtes*. Edited and translated by Heinz Paetzold. Hamburg: Meiner, 1983.

Baumgarten, Alexander G. *Ästhetik*. Edited and translated by Dagmar Mirbach. Hamburg: Meiner, 2007.

Bedaux, Baptist, and Brett Cooke, eds. *Sociobiology and the Arts*. Atlanta: Rodopi, 1999.

Belt, Thomas. *The Naturalist in Nicaragua*. Chicago: University of Chicago Press, 1985.

Benjamin, Walter. "Goethe's Elective Affinities." In *Walter Benjamin: Selected Writings, vol.1, 1913–1926*, edited by Marcus Bullock and Michael W. Jennings, 297–360. Cambridge, MA: The Belknap Press of Harvard University Press, 1999.

Benjamin, Walter. *The Arcades Project*. Translated by Howard Eiland and Kevin McLaughlin. Cambridge, MA: The Belknap Press of Harvard University Press, 1999.

Benthien, Claudia. *Haut. Literaturgeschichte – Körperbilder – Grenzdiskurse*. Reinbek: Rowohlt, 1999.

Benzon, William L. *Beethoven's Anvil: Music in Mind and Culture*. Oxford: Oxford University Press, 2002.

Benzon, William L. "The Evolution of Narrative and the Self." *Journal of Social and Evolutionary Systems* 16 (1993): 129–55. Accessed August 10, 2017. doi:10.1016/1061-7361(93)90025-M.

Berlyne, Dennis E. *Aesthetics and Psychobiology*. New York: Appelton-Century-Crofts, 1971.

Berlyne, Daniel E., ed. *Studies in the New Experimental Aesthetics: Steps Toward an Objective Psychology of Aesthetic Appreciation*. Washington, DC: Hemisphere Publishing Corporation, distributed by Wiley, 1974.

Blacking, John. *How Musical Is Man?* Seattle: University of Washington Press, 1973.

Blaffer-Hrdy, Sarah. *Mothers and Others. The Evolutionary Origins of Mutual Understanding*. Cambridge, MA: The Belknap Press of Harvard University Press, 2009.

Boesch, Christophe, and Hedwige Boesch. "Tool Use and Tool Making in Wild Chimpanzees." *Folia Primatologica* 54 (1990): 86–99.

Bohrn, Isabel C., Ulrike Altmann, Oliver Lubrich, Winfried Menninghaus, and Arthur M. Jacobs. "When We Like What We Know—A Parametric FMRI Analysis of Beauty and familiarity." *Brain and Language* 124, no. 1 (2013): 1–8. Accessed August 22, 2017. doi:10.1016/j.bandl.2012.10.003.

Borbolla de la, Rubín. “Types of Tooth Mutilations Found in Mexico.” *American Journal of Physical Anthropology* 26 (1940): 359–65. Accessed August 10, 2017. doi:10.1002/ajpa.1330260136.

Bornstein, Robert F. “Exposure and Affect: Overview and Meta-Analysis of Research 1968–1987.” *Psychological Bulletin* 106 (1989): 265–89.

Bourdieu, Pierre. *Distinction: A Social Critique of the Judgement of Taste.* Translated by Richard Nice. Cambridge, MA: Harvard University Press, 1984.

Bouzouggar, Abdeljalil, Nick Barton, Marian Vanhaeren, Francesco d’Errico, Simon Collcutt, Tom Higham, Edward Hodge, Simon Parfitt, Edward Rhodes, Jean-Luc Schwenninger, Chris Stringer, Elaine Turner, Steven Ward, Abdelkrim Moutmir, and Abdelhamid Stambouli. “82,000-Year-Old Shell Beads from North Africa and Implications for the Origins of Modern Human Behavior.” *Proceedings of the National Academy of Sciences* 104, no. 24 (2007): 9964–69.

Boyd, Brian. “Evolutionary Theories of Art.” In *The Literary Animal: Evolution and the Nature of Narrative,* edited by Jonathan Gottschall and David S. Wilson, 147–76. Evanston, IL: Northwestern University Press, 2005.

Boyd, Brian. *On the Origins of Stories: Evolution, Cognition, and Fiction.* Cambridge, MA: The Belknap Press of Harvard University Press, 2009.

Boyd, Robert, and Peter J. Richerson. *Culture and the Evolutionary Process.* Chicago: University of Chicago Press, 1985.

Boyer, Pascal. “Religious Thought and Behaviour as By-Products of Brain Function.” *Trends in Cognitive Science* 7 (2003): 119–24.

Boyer, Pascal, and Ramble, Charles. “Cognitive Templates for Religious Concepts: Cross-Cultural Evidence for Recall of Counter-Intuitive Representations.” *Cognitive Science* 25 (2001): 535–64. Accessed August 10, 2017. doi:10.1207/s15516709cog2504_2.

Bråten, Stein, ed. *Intersubjective Communication and Emotion in Early Ontogeny.* Cambridge: Cambridge University Press, 1998.

Bredekamp, Horst. *Darwins Korallen. Die frühen Evolutionsdiagramme und die Tradition der Naturgeschichte.* Berlin: Klaus Wagenbach, 2005.

Bredekamp, Horst. *Theorie des Bildakts - Frankfurter Adorno-Vorlesungen 2007.* Berlin: Suhrkamp, 2010.

Brewer, William F., and Glenn V. Nakamura. “The Nature and Functions of Schema.” *Technical Report,* no. 325 (1984). Accessed July 7, 2011. http://www.eric.ed.gov/PDFS/ED248491.pdf.

Brown, Donald E. *Human Universals.* Philadelphia: Temple University Press, 1991.

Brown, Steven, Björn Merker, and Nils L. Wallin. “An Introduction to Evolutionary Musicology.” In *The Origins of Music,* edited by Nils L. Wallin, Björn Merker, and Steven Brown, 3–24. Cambridge, MA: The MIT Press, 2000.

Brown, Steven. “The ‘Musilanguage’ Model of Music Evolution.” In *The Origins of Music,* edited by Nils L. Wallin, Björn Merker, and Steven Brown, 271–300. Cambridge, MA: The MIT Press, 2000.

Bruner, Jerome S. *Acts of Meaning*. Cambridge, MA: Harvard University Press, 1990.

Burghardt, Gordon M. *The Genesis of Animal Play: Testing the Limits*. Cambridge, MA: MIT Press, 2005.

Burkart, Judith M., Sarah Balffer-Hrdy, and Carel P. van Schaik. "Cooperative Breeding and Human Cognitive Evolution." *Evolutionary Anthropology* 18 (2009): 175–86.

Burkart, Judith M, and Carel P. van Schaik. "Cognitive Consequences of Cooperative Breeding in Primates?" *Animal Cognition* 13 (2010): 1–19. Accessed August 10, 2017. doi:10.1007/s10071-009-0263-7.

Burke, Edmund. A Philosophical Enquiry into the Origins of Our Ideas of the Sublime and the Beautiful. Edited by James T. Boulton. Notre Dame, IN: University of Notre Dame Press, 1968.

Burkert, Walter. *Anthropologie des religiösen Opfers. Die Sakralisierung der Gewalt*. München: Carl-Friedrich-von-Siemens-Stiftung, 1987.

Burley, Nancy. "Sex Ratio Manipulation and Selection for Attractiveness." *Science* 211, no. 4483 (1981): 721–22.

Burley, Nancy. "Leg-Band Color and Mortality Patterns in Captive Breeding Populations of Zebra Finches." *The Auk* 102, no. 3 (1985): 647–51.

Burley, Nancy. "Sexual Selection for Aesthetic Traits in Species with Biparental Care." *American Naturalist* 127, no. 4 (1986): 415–45.

Burley, Nancy. "Comparison of the Band-Colour Preferences of Two Species of Estrildid Finches." *Animal Behaviour* 34, no. 6 (1986): 1732–41. Accessed August 10, 2017. doi: 10.1016/S0003-3472(86)80260-3.

Burley, Nancy. "Sex-Ratio Manipulation in Color-Banded Populations of Zebra Finches." *Evolution* 40, no. 6 (1986): 1191–1206. Accessed September, 28, 2017. doi: 10.2307/2408947.

Burley, Nancy. "Wild Zebra Finches Have Band-Colour Preferences." *Animal Behaviour* 36, no. 4 (1988): 1235–37. Accessed September 9, 2017. doi:10.1016/S0003-3472(88)80085-X.

Burley, Nancy. "The Differential-Allocation Hypothesis: An Experimental Test." *American Naturalist* 132, no. 5 (1988): 611–28.

Busch, Werner. *Die notwendige Arabeske. Wirklichkeitsaneignung und Stilisierung in der deutschen Kunst des 19. Jahrhunderts*. Berlin: Mann, 1985.

Buss, David M. "Sex Differences in Human Mate Selection Preferences: Evolutionary Hypotheses Tested in 37 Cultures." *Behavioral and Brain Sciences* 12, no. 1 (2010): 1–14. Accessed August 4, 2017. doi:10.1017/S0140525X00023992.

Bygren, Lars, Boinkum B. Konlaan, and Sven-E. Johansson. "Attendance at Cultural Events, Reading Books or Periodicals, and Making Music or Singing in a Choir as Determinants for Survival: Swedish Interview Survey of Living Conditions." *British Medical Journal* 313, no. 7072 (1996): 1577–80.

Carrithers, Michael. "Narrativity: Mindreading and Making Societies." In *Natural Theories of Mind. Evolution, Development and Simulation of Everyday Mindreading*, edited by Andrew Whiten, 305–17. Oxford: Basil Blackwell, 1991.

Carroll, Joseph. *Literary Darwinism. Evolution, Human Nature, and Literature*. New York: Routledge, 2004.

Catchpole, Clive K., and Peter J. Slater. *Birdsong: Biological Themes and Variations*. Cambridge: Cambridge University Press, 2008.

Catterall, James S. "Does Experience in the Arts Boost Academic Achievement? A Response to Eisner." *Art Education* 51, no. 4 (1998): 6–11. Accessed August 7, 2017. doi:10.1080/00043125.1998.11654333.

Caterall, James S., Richard Chapleau, and John Iwanaga. *Champions of Change: The Impact of the Arts on Learning*, edited by Edward B. Fiske, 1–19. Washington, DC: Arts Education Partnership, President's Committee on the Arts and the Humanities, GE Fund, and The John D. and Catherine T. MacArthur Foundation, 1999. Accessed July 7, 2011. http://artsedge.kennedy-center.org/champions/pdfs/ChampsReport.pdf.

Cavalli-Sforza, Luigi L., and Marcus. W. Feldman. "Cultural versus Biological Inheritance: Phenotypic Transmission from Parents to Children. (A Theory of the Effect of Parental Phenotypes on Children's Phenotypes)." *American Journal of Human Genetics* 25, no. 6 (1973): 618–37.

Chatterjee, Anjan, Amy Thomas, Sabrina E. Smith, and Geoffrey K. Aguirre. "The Neural Response to Facial Attractiveness." *Neuropsychology* 23, no. 2 (2009): 135–43. Accessed August 21, 2017. doi:10.1037/a0014430.

Cheney, Dorothy L., and Robert M. Seyfarth. *How Monkeys See the World: Inside the Mind of Another Species*. Chicago: The University of Chicago Press, 1992.

Cheney, Dorothy L., and Robert M. Seyfarth. "Primate Communication and Human Language: Continuities and Discontinuities." In *Mind the Gap. Tracing the Origin of Human Universals*, edited by Peter M. Kappeler and Joan B. Silk, 283–98. Berlin: Springer, 2010.

Chipp, Herschel B. "Formal and Symbolic Factors in the Art Styles of Primitive Cultures." In *Art and Aesthetics in Primitive Societies: A Critical Anthology*, edited by Carol F. Jopling, 146–70, New York: E. P. Dutton & Co., 1971.

Coe, Kathryn. "Art: The Replicable Unit—An Inquiry into the Possible Origin of Art as a Social Behavior." *Journal of Social and Evolutionary Systems* 15, no. 2 (1992): 217–34. Accessed September 6, 2017. doi:10.1016/1061-7361(92)90005-X.

Coe, Kathryn. *The Ancestress Hypothesis*. New Brunswick, NJ: Rutgers University Press, 2003.

Coleridge, Samuel Taylor. "Biographia Literaria, Volume II." In *The Collected Works of Samuel Taylor Coleridge: Biographia Literaria, or Biographical Sketches of My Literary Life and Opinions*, edited by James Engell and Walter Jackson Bate. Princeton, NJ: Princeton University Press, 1983.

Comerford Boyes, Louise, and Ivan Reid. "What are the Benefits for Pupils Participating in Arts Activities?" *Research in Education* 73, no. 1 (2005): 1–14. Accessed August 14, 2017. doi:10.7227/RIE.73.1.

Conard, Nicholas J., and Bolus, Michael. "Radiocarbon Dating the Appearance of Modern Humans and Timing of Cultural Innovations in Europe: New Results and New Challenges." *Journal of Human Evolution* 44, (2003): 331–71. Accessed August 10, 2017. doi:10.1016/S0047-2484(02)00202-6.

Conard, Nicholas J., Maria Malina, Susanne C. Münzel, and Friedrich Seeberger. "Eine Mammutelfenbeinflöte aus dem Aurignacien des Geissenklösterle." *Archäologisches Korrespondenzblatt* 34, no. 4 (2004): 447–62.

Conard, Nicholas J., Michael Bolus, Paul Goldberg, and Susanne C. Münzel. "The Last Neanderthals and First Modern Humans in the Swabian Jura." In *When Neanderthals and Modern Humans Met,* edited by Nicholas J. Conard, 305–41. Tübingen: Kerns, 2006.

Conard, Nicholas J. "Cultural Evolution in Africa and Eurasia during the Middle and Late Pleistocene." In *Handbook of Paleoanthropology*, edited by Winfried Henke, Ian Tattersall, and Hardt Thorolf, 2001–37. Berlin, Heidelberg u.a.: Springer-Verlag, 2007.

Conard, Nicolas J. "A Critical View of the Evidence for a Southern African Origin of Behavioural Modernity." *South African Archaeological Society Goodwin Series* 10, (2008): 175–79.

Conard, Nicholas J. "A Female Figurine from the Basal Aurignacian of Hohle Fels Cave in Southwestern Germany." *Nature* 459 (2009): 248–52. Accessed August 10, 2017. doi:10.1038/nature07995.

Conard, Nicholas J. "Die erste Venus. Zur ältesten Frauendarstellung der Welt." In *Eiszeit: Kunst und Kultur*, edited by Susanne Rau and Archäologisches Landesmuseum Baden-Württemberg, 268–71. Ostfildern: Thorbecke, 2009.

Conard, Nicholas J., and Petra Kieselbach. "Eindeutig männlich! Ein Phallus aus dem Hohle Fels." In *Eiszeit: Kunst und Kultur*, edited by Susanne Rau and Archäologisches Landesmuseum Baden-Württemberg, 282–86. Ostfildern: Thorbecke, 2009.

Conard, Nicholas J., Maria Malina, and Susanne C. Münzel. "New Flutes Document the Earliest Musical Tradition in Southwestern Germany." *Nature* 460 (2009): 737–40. Accessed August 10, 2017. doi:10.1038/nature08169.

Conkey, Margaret W. "New Approaches in the Search for Meaning? A Review of Research in 'Paleolithic Art." *Journal of Field Archaeology* 14, no. 4 (1987): 413–30. Accessed August 04, 2017. doi:10.1179/jfa.1987.14.4.413.

Cooke, Brett. "Edward O. Wilson on Art." In *Biopoetics. Evolutionary Explorations in the Arts*, edited by Brett Cooke and Frederick Turner, 97–118. Lexington: ICUS, 1999.

Coote, Jerem, and Anthony Shelton. *Anthropology, Art, and Aesthetics.* Oxford: Clarendon Press, 1992.

Cosmides, Leda, and John Tooby. "Consider the Source: The Evolution of Adaptations for Decoupling and Metarepresentation." In *Metarepresentations. A Multidisciplinary Perspective*, edited by Dan Sperber, 53–116. Oxford: Oxford University Press, 2000.

Cramer, Friedrich. *Chaos und Ordnung. Die komplexe Struktur des Lebendigen.* Stuttgart: Deutsche Verlags-Anstalt. 1988.

Crockford, Catherine, Ilka Herbinger, Linda Vigilant, and Christophe Boech. "Wild Chimpanzees Produce Group-Specific Calls: A Case for Vocal Learning." *Ethology* 110 (2004): 221–43. Accessed August 10, 2017. doi:10.1111/j.1439-0310.2004.00968.x.

Cronin, Helena. *The Ant and the Peacock: Altruism and Sexual Selection from Darwin to Today.* Cambridge: Cambridge University Press, 1991.

Cross, Ian. "The Nature of Music and Its Evolution." In *The Oxford Handbook of Music Psychology*, edited by Susan Hallam, Ian Cross, and Michael Thaut, 3–13. Oxford,: Oxford University Press, 2009.

Culham, Jody C., and Kenneth F. Valyear. "Human Parietal Cortex in Action." *Current Opinion in Neurobiology* 16, no. 2 (2006): 205–12. Accessed August 10, 2017. doi:10.1016/j.conb.2006.03.005.

Cupchik, Gerald C., Garry Leonard, Elise Axelrad, and Judith D. Kalin. "The Landscape of Emotion in Literary Encounters." *Cognition and Emotion* 12, no. 6 (1998): 825–47.

Cupchik, Gerald C., Oshin Vartanian, Adrian Crawley, and David J. Mikulis. "Viewing Artworks: Contributions of Cognitive Control and Perceptual Facilitation to Aesthetic Experience." *Brain and Cognition* 70, no. 1 (2009): 84–91. Accessed October 1, 2014. doi:10.1016/j.bandc.2009.01.003.

d'Errico, Francesco, and Marid Soressi. "Systematic Use of Pigment by Pech-de-l'Azé Neandertals: Implications for the Origin of Behavioural Modernity." *Journal of Human Evolution* 42 (2002): 13.

d'Errico, Francesco. "The Invisible Frontier. A Multiple Species Model for the Origin of Behavioral Modernity." *Evolutionary Anthropology: Issues, News, and Reviews* 12, no. 4 (2003): 188–202. Accessed August 10, 2017. doi:10.1002/evan.10113.

d'Errico, Francesco, Christopher Henshilwood, Marian Vanhaeren, and Karen van Niekerk. "Nassarius Kraussianus Shell Beads from Blombos Cave: Evidence for Symbolic Behaviour in the Middle Stone Age." *Journal of Human Evolution* 48, no. 1 (2005): 3–24. Accessed August 10, 2017. doi:10.1016/j.jhevol.2004.09.002.

d'Errico, Francesco, Marian Vanhaeren, Nick Barton, Abdeljalil Bouzouggar, Henk Mienis, Daniel Richter, Jean-Jacques Hublin, Shannon P. Mcpherron, and Pierre Lozouet. "Additional Evidence on the Use of Personal Ornaments in the Middle Paleolithic of North Africa." *Proceedings of the National Academy of Sciences of the USA* 106, no. 38 (2009): 16051–56. Accessed September 7, 2017. doi:10.1073/pnas.0903532106.

Derryberry, Elizabeth P. "Evolution of Bird Song Affects Signal Efficacy: An Experimental Test Using Historical and Current Signals." *Evolution* 61, no. 8 (2007): 1938–45. Accessed August 23, 2017. doi:10.1111/j.1558-5646.2007.00154.x.

Darwin, Charles. *The Descent of Man, and Selection in Relation to Sex*. 2nd edition. New York, London: Merril and Baker, 1874.

Darwin, Charles. *The Expression of the Emotions in Man and Animals*. Chicago: University of Chicago Press, 1965.

Darwin, Charles. "Sexual Selection in Relation to Monkeys." In *The Collected Papers of Charles Darwin,* edited by Paul H. Barrett, 285–91. Chicago: Chicago University Press, 1977.

Darwin, Charles. *The Descent of Man, and Selection in Relation to Sex*. Princeton, NJ: Princeton University Press, 1981.

Darwin, Charles. *The Origin of Species*. Oxford: Oxford University Press, 1996.

Darwin, George H. "Development in Dress." *Macmillan's Magazine* 26 (1872): 410.

Davies, Stephen. *The Artful Species: Aesthetics, Art, and Evolution*. Oxford: Oxford University Press, 2012.

Dawkins, Richard. *The Selfish Gene*. Oxford: Oxford University Press, 2006.

Deacon, Terrence. "The Aesthetic Faculty." In *The Artful Mind: Cognitive Science and the Riddle of Human Creativity*, edited by Mark Turner, 21–53. Oxford: Oxford University Press, 2006.

Deasy, Richard J., ed. "Critical Links: Learning in the Arts and Student Academic and Social Development." Washington, DC: Arts Education Partnership. 2002.

Dediu, Dan, and Stephen Levinson. "On the Antiquity of Language: The Reinterpretation of Neandertal Linguistic Capacities and its Consequences." *Frontiers in Psychology* 4, no. 397 (2013). Accessed July 26, 2017. doi:10.3389/fpsyg.2013.00397.

Defleur, Alban. *Les Sépultures Moustériennes*. Paris: CNRS Editions, 1993.

Dehaene, Stanislas, Laurent Cohen, Mariano Sigman, and Fabien Vinckier. "The Neural Code for Written Words: A Proposal." *Trends in Cognitive Sciences* 9, no. 7 (2005): 335–41. Accessed July 24, 2017. doi:10.1016/j.tics.2005.05.004.

Dehaene, Stanislas, and Laurent Cohen. "Cultural Recycling of Cortical Maps." *Neuron* 56, no. 2 (2007): 384–98. Accessed July 24, 2017. doi:10.1016/j.neuron.2007.10.004.

Dempster, Douglas. "Is There Even a Grammar of Music?" *Musicae Scientiae* 2, no. 1 (1998): 55–65. Accessed July 24, 2017. doi:10.1177/102986499800200104.

Dermer, Marshall, and Darrel L. Thiel. "When Beauty May Fail." *Journal of Personality and Social Psychology* 31, no. 6 (1975): 1168–76.

Devereux, George. "Art and Mythology: A General Theory." In *Art and Aesthetics in Primitive Societies: A Critical Anthology*, edited by Carol F. Jopling, 193–224. New York: E. P. Dutton & Co., 1971.

de Waal, Frans B. M. *Good Natured: The Origins of Right and Wrong in Humans and Other Animals*. Cambridge, MA: Harvard University Press, 1997.

de Waal, Frans B. M. *The Age of Empathy: Nature's Lessons for a Kinder Society*. New York: Harmony Books, 2009.

Diener, Ed, Brian Wolsic, and Frank Fujita. "Physical Attractiveness and Subjective Well-Being." *Journal of Personality and Social Psychology* 69, no. 1 (1995): 120–29. Accessed August 14, 2017. doi:10.1037/0022-3514.69.1.120.

Dijkstra, Katinka, Rolf A. Zwaan, Arthur C. Graesser, and Joseph P. Magliano. "Character and Reader Emotions in Literary Texts." *Poetics* 23, no. 1 (1995): 139–57. Accessed August 14, 2017. doi:10.1016/0304-422X(94)00009-U.

DiMaggio, Paul. "Cultural Capital and School Success: The Impact of Status Culture Participation on the Grades of U.S. High School Students." *American Sociological Review* 47, no. 2 (1982): 189–201.

Dingwall, Eric J. *Artificial Cranial Deformation: A Contribution to the Study of Ethnic Mutilations*. London: John Bale, Sons and Danielsson, 1931.

Dissanayake, Ellen. *What Is Art For?* Seattle: University of Washington Press, 1988.

Dissanayake, Ellen. *Homo Aestheticus. Where Art Comes from and Why*. Seattle: University of Washington Press, 1999.

Dissanayake, Ellen. *Art and Intimacy. How the Arts Began*. Seattle: University of Washington Press, 2000.

Dissanayake, Ellen. "Antecedents of the Temporal Arts in Early Mother-Infant Interaction." In *The Origins of Music*, edited by Nils L. Wallin, Björn Merker, and Steven Brown, 389–410. Cambridge, MA: The MIT Press, 2000.

Dissanayake, Ellen. "The Arts after Darwin: Does Art Have an Origin and Adaptive Function?" In *World Art Studies: Exploring Concepts and Approaches*, edited by Kitty Zijlmans and Wilfried van Damme, 241–63. Amsterdam: Valiz, 2008.

Dissanayake, Ellen. "Art as a Human Universal: An Adaptationist View." In *Origins of Religion, Cognition and Culture*, edited by Armin W. Geertz, 121–39. Durham: Acumen Publishing Limited, 2013.

Dixson, Barnaby J., Alan F. Dixson, Phil J. Bishop, and Amy Parish. "Human Physique and Sexual Attractiveness in Men and Women: A New Zealand-U.S. Comparative Study." Archives of Sexual Behavior 39, no. 3 (2010): 798–806.

Donald, Merlin. "Art and Cognitive Evolution." In *The Artful Mind, Cognitive Science and the Riddle of Creativity*, edited by Mark Turner, 3–20. Oxford: Oxford University Press, 2006.

Dunbar, Robin. *Grooming, Gossip and the Evolution of Language*. London: Faber & Faber, 1996.

Dunbar, Robin. "The Social Brain Hypothesis." *Evolutionary Anthropology* 6, no. 5 (1998): 178–90.

Dunbar, Robin. *The Human Story. A New History of Mankind's Evolution*. London: Faber & Faber, 2004.

Dunbar, Robin, and Susanne Shultz. "Evolution in the Social Brain." *Science* 317, no. 5843 (2007): 1344. Accessed September 4, 2017. doi:10.1126/science.1145463.

Dunbar, Robin. "Coevolution of Neocortical Size, Group Size and Language in Humans." *Behavioral and Brain Sciences* 16, no. 4 (2010): 681–94. Accessed August 10, 2017. doi:10.1017/S0140525X00032325.

Dunn, Michael, Angela Terrill, Ger Reesink, Robert A. Foley, and Stephen C. Levinson. "Structural Phylogenetics and the Reconstruction of Ancient Language History." *Science* 309, no. 5743 (2005): 2072. Accessed July 26, 2017. doi:10.1126/science.1114615.

Dunn, Michael, Stephen Levinson, Eva Lindström, Ger Reesink, and Angela Terrill. "Structural Phylogeny in Historical Linguistics: Methodological Explorations Applied in Island Melanesia." *Language* 84, no. 4 (2008): 710–59.

Dunn, Michael, Simon J. Greenhill, Stephen J. Levinson, and Russell D. Gray. "Evolved Structure of Languages Shows Lineage-Specific Trends in Word-Order Universals." *Nature* 473, no. 7345 (2011): 79–82. Accessed July 26, 2017. doi:10.1038/nature09923.

Durkheim, Émile. *The Elementary Forms of the Religious Life*. Translated by Joseph Ward Swain. Mineola, NY: Dover Publications. 2008.

Dutton, Dennis. *The Art Instinct. Beauty, Pleasure and Human Evolution*. New York: Bloomsbury, 2009.

Eagly, Alice H., Richard D. Ashmore, Mona G. Makhijani, and Laura C Longo. "What Is Beautiful Is Good, but . . . : A Meta-Analytic Review of Research on the Physical Attractiveness Stereotype." *Psychological Bulletin* 110, no. 1 (1991): 109–28. Accessed September 28, 2017. doi:10.1037/0033-2909.110.1.109.

Eberhard, William G. *Sexual Selection and Animal Genitalia*. Cambridge, MA: Harvard University Press, 1985.

Ehrenreich, Barbara, Elisabeth Hess, and Gloria Jacobs. "Beatlemania. Girls just Want to Have Fun." In *Re-Making Love. The Feminization of Sex*, edited by Barbara Ehrenreich, Elizabeth Hess and Gloria Jacobs, 10–38. Garden City, NY: Doubleday, 1986.

Eibl, Karl. *Animal Poeta. Bausteine der biologischen Kultur- und Literaturtheorie*. Paderborn: Mentis, 2004.

Eibl, Karl. *Kultur als Zwischenwelt. Eine evolutionsbiologische Perspektive*. Frankfurt am Main: Suhrkamp, 2009.

Eibl-Eibesfeldt, Irenäus. "The Biological Foundation of Aesthetics." In *Beauty and the Brain: Biological Aspects of Aesthetics*, edited by Ingo Rentschler, Barbara Herzberger and David Epstein, 29–68. Boston: Birkhäuser, 1988.

Eibl-Eibesfeldt, Irenäus. "Ernst Haeckel—The Artist in the Scientist." In *Art Forms in Nature: The Prints of Ernst Haeckel*, edited by Ernst Haeckel, Olaf Breidbach, and Irenäus Eibl-Eibesfeldt, translated by Michele Schons, 19–30. New York: Prestel, 2008.

Eisner, Elliot W. "Does Experience in the Arts Boost Academic Achievement?" *The Clearing House* 72, no. 3 (1999): 143–49.

Eisner, Elliot W. "What Justifies Arts Education: What Research Does Not Say." In *Enlightened Advocacy: Implications of Research for Arts Education Policy Practice*, Proceedings of the 1999 Charles Fowler Colloquium on Innovation in Arts Education, edited by Marie McCarthy, 19–29. College Park: University of Maryland, 2000.

Elowson, A. Margaret, Charles T. Snowdon, and Cristina Lazaro-Perea. "Babbling' and Social Context in Infant Monkeys: Parallels to Human Infants." *Trends in Cognitive Sciences* 2, no. 1 (1998): 31–37. Accessed August 14, 2017. doi:10.1016/S1364-6613(97)01115-7.

Endler, John A., Lorna C. Endler, and Natalie R. Doerr. "Great Bowerbirds Create Theaters with Forced Perspective when Seen by Their Audience." *Current Biology* 20, no. 18 (2010): 1679–84. Accessed August 14, 2017. doi:10.1016/j.cub.2010.08.033.

Etcoff, Nancy. *Survival of the Prettiest. The Science of Beauty.* New York: Doubleday, 1999.

Fagen, Robert M. *Animal Play Behavior.* Oxford: Oxford University Press, 1981.

Falk, Dean. "Hominid Brain Evolution and the Origins of Music." In *The Origins of Music*, edited by Nils L. Walin, Björn Merker, and Steven Brown, 197-216. Cambridge, MA: MIT Press, 2001.

Falk, Dean. "Prelinguistic Evolution in Early Hominids: Whence Motherese?" *Behavioral and Brain Sciences* 27, no. 4 (2005): 491–503. Accessed July 10, 2017. doi:10.1017/S0140525X04000111.

Fechner, Gustav T. *Vorschule der Ästhetik*. Leipzig: Breitkopf & Härtel, 1876.

Feldman, Marcus W., and Luigi L. Cavalli-Sforza. "Cultural and Biological Evolutionary Processes: Selection for a Trait under Complex Transmission." *Theoretical Population Biology* 9 (1976): 238–59.

Feldman, Marcus W., and Lev A. Zhivotovsky. "Gene-Culture Coevolution: Toward a General Theory of Vertical Transmission." *Proceedings of the National Academy of Sciences. USA* 89 (1992): 11935–38.

Feldman, Marcus W., and Kevin N. Laland. "Gene-Culture Coevolutionary Theory." *Trends in Ecology and Evolution* 11, no. 11 (1996): 453–57. Accessed August 14, 2017. doi:10.1016/0169-5347(96)10052-5.

Feuk, Lars, Aino Kalervo, Marita Lipsanen-Nyman, Jennifer Skaug, and Brenda F. Nakabayashi. "Absence of a Paternally Inherited FOXP2 Gene in Developmental Verbal Dyspraxia." *American Journal of Human Genetics* 79, no. 5 (2006): 965–72.

Fischer, John L. "Art Styles as Cultural Cognitive Maps." In *Art and Aesthetics in Primitive Societies: A Critical Anthology*, edited by Carol F. Jopling, 171–92. New York: E. P. Dutton & Co., 1971.

Fisher, Ronald Aylmer. *The Genetical Theory of Natural Selection*. Oxford: Clarendon, 1930.

Fisher, Simon E., and Marcus, Gary F. "The Eloquent Ape: Genes, Brains and the Evolution of Language." *Nature Reviews. Genetics* 7, no. 1 (2006): 9–20.

Fiske, Edward B, ed. *Champions of Change. The Impact of the Arts on Learning*, Washington: Arts Education Partnership, 1999. Accessed July 7, 2011. http://artsedge.kennedy-center.org/champions/ pdfs/ChampsReport.pdf.

Fitch, Tecumseh W. "Dancing to Darwin's Tune." *Nature* 438 (2005): 288.

Fitch, Tecumseh W. "The Biology and Evolution of Music: A Comparative Perspective." *Cognition* 100, no. 1 (2006): 173–215. Accessed August 14, 2017. doi:10.1016/j.cognition.2005.11.009.

Fitch, Tecumseh W. *The Evolution of Language*. New York: Cambridge University Press, 2010.

Floss, Harald, and Nathalie Rouquerol, eds. *Les Chemin de l'Art Aurignacien en Europe. Das Aurignacien und die Anfänge der Kunst in Europa*. Aurignac: Éditions Musée-forum Aurignac, Cahier 4, 2007.

Floss, Harald. "Die frühesten Bildwerke der Menschheit: Das Phänomen Eiszeitkunst." In *Eiszeit: Kunst und Kultur*, edited by Susanne Rau and Archäologisches Landesmuseum Baden-Württemberg, 228–41. Ostfildern: Thorbecke, 2009.

Fodor, Jerry A. *The Modularity of Mind: An Essay on Faculty Psychology*. Cambridge, MA: The MIT Press, 1983.

Fredrickson, Barbara L. "The Role of Positive Emotions in Positive Psychology: The Broaden-and-Build Theory of Positive Emotions." *American Psychologist* 56, no. 3 (2001): 218–26.

Freedman, Walter. "A Neurobiological Role of Music in Social Bonding." In *The Origins of Music*, edited by Nils L. Wallin, Björn Merker and Steven Brown, 411–24. Cambridge, MA: The MIT Press, 2000.

Freud, Sigmund. "Three Essays on the Theory of Sexuality." In *The Standard Edition of the Complete Psychological Works of Sigmund Freud*, translated and edited by James Strachey, in collaboration with Anna Freud, 125–245. Vol. 7. London: Hogarth Press, 1953.

Freud, Sigmund. "Totem and Taboo." In *The Standard Edition of the Complete Psychological Works of Sigmund Freud.* Translated and edited by James Strachey, in collaboration with Anna Freud, 1–162. Vol. 13. London: Hogarth Press, 1955.

Freud, Sigmund. "Civilization and its Discontents." In *The Standard Edition of the Complete Psychological Works of Sigmund Freud.* Translated and edited by James Strachey, in collaboration with Anna Freud, 59–145. Vol. 21. London: Hogarth Press, 2001.

Freud, Sigmund. "The 'Uncanny'." In *The Standard Edition of the Complete Psychological Works of Sigmund Freud.* Translated and edited by James Strachey, in collaboration with Anna Freud, 217–56. Vol. 17. London: Hogarth Press, 2001.

Freud, Sigmund. "The Future of an Illusion." In *The Standard Edition of the Complete Psychological Works of Sigmund Freud.* Translated and edited by James Strachey, in collaboration with Anna Freud, 1–56. Vol. 21. London: Hogarth Press, 2001.

Frijda, Nico H. "The Laws of Emotion." *American Psychologist* 43, no. 5 (1988): 349–58.

Frijda, Nico H. and Louise Sundararajan. "Emotion Refinement: A Theory Inspired by Chinese Poetics." *Perspectives on Psychological Science* 2, no. 3 (2007): 227–41. Accessed August 10, 2017. doi: 10.1111/j.1745-6916.2007.00042.x.

Frisch, Karl von. *Animal Architecture.* New York: Harcourt Brace Jovanovich, 1974.

Friston, Karl. "The Free-Energy Principle: a Rough Guide to the Brain?" *Trends in Cognitive Sciences* 13, no. 7 (2009): 293–301. Accessed August 10, 2017. doi:10.1016/j.tics.2009.04.005.

Fritz, Thomas, Sebastian. Jentschke, Nathalie Gosselin, Daniela Sammler, Isabelle Peretz, Robert Turner, Angeala D. Friederici, and Stefan Koelsch. "Universal Recognition of Three Basic Emotions in Music." *Current Biology* 19, no. 7 (2009): 573–76. Accessed July 25, 2017. doi:10.1016/j.cub.2009.02.058.

Fritzsche, Bettina. *Pop-Fans. Studie einer Mädchenkultur.* Opladen: Leske und Buderich, 2003.

Fritzsche, Bettina. "Fans und Gender." In *Fans. Soziologische Perspektiven*, edited by Jochen Roose Jochen, Mike S. Schäfer and Thomas Schmidt-Lux, 229–48. Wiesbaden: VS Verlag für Sozialwissenschaften, 2010.

Füller, Horst. *Die Schönheit der Tiere. Studien über die tierische Erscheinung.* Leipzig: Urania-Verlag, 1995.

Gangestad, Steven W. "Sexual Selection and Physical Attractiveness: Implications for Mating Dynamics." *Human Nature* 4, no. 3 (1993): 205–35. Accessed July 24, 2017. doi:10.1007/BF02692200.

Gangestad, Steven W., and David M. Buss. "Pathogen Prevalence and Human Mate Preferences." *Ethology and Social Biology* 14, no. 2 (1993): 89–96. Accessed July 24, 2017. doi:10.1016/0162-3095(93)90009-7.

Gans, Eric. *Originary Thinking. Elements of Generative Anthropology.* Stanford, CA: Stanford University Press, 1993.

Gat, Azar. "Why War? Motivations for Fighting in the Human State of Nature." In *Mind the Gap. Tracing the Origin of Human Universals*, edited by Peter M. Kappeler and Joan B. Silk, 197–220. Berlin: Springer, 2010.

Gaut, Berys. *Art, Emotion and Ethics*. Oxford: Oxford University Press, 2007.

Gehlen, Arnold. "Über instinktives Ansprechen auf Wahrnehmungen." In *Anthropologische Forschung. Zur Selbstbegegnung und Selbstentdeckung des Menschen*, 104–26. Reinbek: Rowohlt, 1961.

Geissmann, Thomas. "Gibbon Songs and Human Music from an Evolutionary Perspective." In *The Origins of Music*, edited by Nils L. Wallin, Björn Merker, and Steven Brown, 103–24. Cambridge, MA: The MIT Press, 2000.

Gell, Alfred. "The Technology of Enchantment and the Enchantment of Technology." In *Anthropology, Art, and Aesthetics*, edited by Jeremy Coote and Anthony Shelton, 40–63. Oxford: Clarendon Press, 1992.

Giedion, Siegfried. *The Eternal Present: A Contribution on Constancy and Change*. New York: Pantheon Books, 1962.

Giora, Rachel, and Ofer Fein. "Weapons of Mass Distraction: Optimal Innovation and Pleasure Ratings." *Metaphor and Symbol* 19, no. 2 (2004): 115–41. Accessed August 10, 2017. doi:10.1207/s15327868ms1902_2.

Girard, René. *Violence and the Sacred*. Translated by Patrick Gregory. Baltimore, MD: The Johns Hopkins University Press, 1979.

Glass, Bentley. "Evolution of Hairlessness in Man." *Science* 152, no. 3720 (1966): 294.

Goethe, Johann Wolfgang. *Elective Affinities*. Translated by David Constantine. Oxford: Oxford University Press, 1994.

Goethe, Johann Wolfgang. *Münchner Ausgabe: Sämtliche Werke nach Epochen seines Schaffens*, edited by Karl Richter, Weimarer Klassik 1798–1806. Vol. 6.1. München: C. Hanser, 1985.

Goethe, Johann Wolfgang. "The Rat-Catcher." In *The Poems of Goethe*, translated by Edgar Alfred Bowring, 108. Boston: S. E. Cassino, 1882

Gorgias, "Encomium of Helen." In *The Greek Sophists*, translated by John Dillon and Tania Gergel, 76–84. London: Penguin, 2003.

Gottschall, Jonathan, and David S. Wilson, eds. *The Literary Animal. Evolution and the Nature of Narrative*. Evanston, IL: Northwestern University Press, 2005.

Gottschall, Jonathan. *The Storytelling Animal: How Stories Make Us Human*. Boston: Houghton Mifflin Harcourt, 2012.

Gould, James L., and Carol G. Gould. *Sexual Selection*. New York: Scientific American Library, 1989.

Gould, Stephen J., and Richard C. Lewontin. "The Spandrels of San Marco and the Panglossian Paradigm: A Critique of the Adaptationist Programme." *Proceedings of the Royal Society* 205 (1979): 581–98.

Gould, Stephen J., and Elisabeth S. Vrba. "Exaptation—A Missing Term in the Science of Form." *Paleobiology* 8 (1982): 4–15.

Gould, Stephen J. "Exaptation: A Crucial Tool for Evolutionary Psychology." *Journal of Social Issues* 47, no. 3 (1991): 43–65. Accessed July 24, 2017. doi:10.1111/j.1540-4560.1991.tb01822.x.

Grafen, Alan. "Sexual Selection Unhandicapped by the Fisher Process." *Journal of Theoretical Biology* 144, no. 4 (1990): 473–516.

Graziosi, Paolo. *Die Kunst der Altsteinzeit*. Stuttgart: Kohlhammer, 1956.

Haesler, Sebastian, Christelle Rochefort, Benjamin Georgi, Pawel Licznerski, Pavel Osten and Constance Scharff. "Incomplete and Inaccurate Vocal Imitation after Knockdown of FoxP2 in Songbird Basal Ganglia Nucleus Area X." *PLOS Biology* 5, no. 12 (2007): e321. Accessed August 7, 2017. doi:10.1371/journal.pbio.0050321.

Haesler, Sebastian, and Constance Scharff. "Genes for Tuning up the Vocal Brain: FoxP2 in Human Speech and Birdsong." In *Neuroscience of Birdsong*, edited by Philip H. Zeigler and Peter Marler, 398–406. New York: Cambridge University Press, 2008.

Hagen, Edward H., and Gregory A. Bryant. "Music and Dance as a Coalition Signaling System." *Human Nature* 14, no. 1 (2003): 21–51. Accessed August 7, 2017. doi:10.1007/s12110-003-1015-z.

Hagen, Edward H., and Peter Hammerstein. "Did Neanderthals and Other Early Humans Sing? Seeking the Biological Roots of Music in the Territorial Advertisements of Primates, Lions, Hyenas, and Wolves." *Musicae Scientiae* (Special Issue 2009–10: *Music and Evolution*): 291–320.

Haidle, Miriam N. "Menschenaffen? Affenmenschen? Menschen! Kognition und Sprache im Altpaläolithikum." In *Woher kommt der Mensch?*, edited by Nicholas Conard, 69–97. Tübingen: Attempto-Verlag, 2004.

Haidle, Miriam N. "Wege zur Kunst. Der Mensch als Schöpfer." In *Eiszeit: Kunst und Kultur*, edited by Susanne Rau and Archäologisches Landesmuseum Baden-Württemberg, 242–43. Ostfildern: Thorbecke, 2009.

Hall, Michelle L. "A Review of Hypotheses for the Functions of Avian Duetting." *Behavioral Ecology and Sociobiology* 55, no. 5 (2004): 415–30. Accessed August 7, 2017. doi:10.1007/s00265-003-0741-x.

Halverson, John. "Art for Art's Sake in the Paleolithic." *Current Anthropology* 28, no. 1 (1987): 63–89.

Hamilton, William D. "The Evolution of Altruistic Behavior." *American Naturalist* 97, no. 896 (1963): 354–56.

Hamilton, William D. and Marlene Zuk. "Heritable True Fitness and Bright Birds: A Role for Parasites?" *Science* 218, no. 4570 (1982): 384.

Hannon, Erin E. and Sandra E. Trehub. "Metrical Categories in Infancy and Adulthood." *Psychological Science* 16, no. 1 (2005): 48–55. Accessed August 7, 2017. doi:10.1111/j.0956-7976.2005.00779.x.

Hanich, Julian, Valentin Wagner, Mira Shah, Thomas Jacobsen, and Winfried Menninghaus. "Why We Like to Watch Sad Films. The Pleasure of Being Moved in Aesthetic Experiences." *Psychology of Aesthetics, Creativity, and the Arts* 8, no. 2 (2014): 130–43. Accessed August 7, 2017. doi:10.1037/a0035690.

Hansen, Brian. "The Prehistory of Theatre: A Path with Six Turnings." In *Biopoetics. Evolutionary Explorations in the Arts*, edited by Brett Cooke and Frederick Turner, 347–66. Lexington, KY: ICUS, 1999.

Hardy, Barbara N. *Tellers and Listeners: The Narrative Imagination*. London: Athlone, 1975.

Harris, Paul L. "The Work of the Imagination." In *Natural Theories of Mind. Evolution, Development and Simulation of Everyday Mindreading*, edited by Andrew Whiten, 283–304. Oxford: Basil Blackwell, 1991.

Harvey, Paul H. and Jack W. Bradbury. "Sexual Selection." In *Behavioural Ecology: An Evolutionary Approach*, edited by John R. Krebs and Nicholas B. Davies, 203–33. Oxford: Blackwell Scientific, 1991.

Hatfield, Elaine, and Susan Sprecher. *Mirror, Mirror. The Importance of Looks in Everyday Life*. Albany: SUNY Press, 1986.

Hauser, Marc D. "The Sound and the Fury: Primate Vocalizations as Reflections of Emotion and Thought." In *The Origins of Music*, edited by Nils L. Wallin, Björn Merker, and Steven Brown, 77–102. Cambridge, MA: The MIT Press, 2000.

Hayles, N. Katherine. *Chaos Bound: Orderly Disorder in Contemporary Literature and Science*. Ithaca, NY: Cornell University Press, 1990.

Heath, Shirley Brice, and Elizabeth Soep. "Youth Development and the Arts in Nonschool Hours." *Grantmakers in the Arts Newsletter* 9, no. 1 (1998): 9–32.

Heeschen, Volker. "The Narration 'Instinct': Everyday Talk and Aesthetic Forms of Communication (in Communities of the New Guinean Mountains)." In *Verbal Art across Cultures: The Aesthetics and Proto-Aesthetics of Communication*, edited by Hubert Knoblauch and Helga Kotthoff, 137–66. Tübingen: GNV Gunter Narr Verlag, 2001.

Henshilwood, Christopher S., Francesco Errico, Royden Yates, Zenobia Jacobs, Chantal Tribolo, Geoff A. T. Duller, Norbert Mercier, Judith C. Sealy, Helene Valladas, Ian Watts, and Ann G. Wintle. "Emergence of Modern Human Behavior: Middle Stone Age Engravings from South Africa." *Science* 295, no. 5558 (2002): 1278. Accessed August 24, 2017. doi:10.1126/science.1067575.

Henss, Ronald. "Waist-to-Hip Ratio and Attractiveness. Replication and Extension." *Personality and Individual Differences* 19, no. 4 (1995): 479–88. Accessed July 19, 2017. doi:10.1016/0191-8869(95)00093-L.

Herder, Johann Gottfried: "Plastik. Einige Wahrnehmungen über Form und Gestalt aus Pygmalions bildendem Träume." In *Sämtliche Werke*, edited by Bernhard Suphan, 1–87. Vol. 8. Berlin: Weidmannsche Buchhandlung. 1892.

Herder, Johann Gottfried. "Studien und Entwürfe zur Plastik." In *Sämtliche Werke*, edited by Bernhard Suphan, 88-115. Vol. 8. Berlin: Weidmannsche Buchhandlung. 1892.

Herder, Johann Gottfried. "*Abhandlung über den Ursprung der Sprache*." In *Frühe Schriften 1764-1772*, edited by Ulrich Gaier, 695–810. Vol. 1. Frankfurt am Main: Deutscher Klassiker Verlag, 1985.

Hetland, Lois, and Ellen Winner. "The Arts and Academic Achievement: What the Evidence Shows: Executive Summary." *Arts Education Policy Review* 102, no. 5 (2001): 3–6. Accessed July 19, 2017. doi:10.1080/10632910109600008.

Heyes, Cecilia. "Grist and Mills: On the Cultural Origins of Cultural Learning." *Philosophical Transactions of the Royal Society B: Biological Sciences* 367, no. 1599 (2012): 2181–91. Accessed July 19, 2017. doi:10.1098/rstb.2012.0120.

Hjort, Mette, and Sue Laver, eds. *Emotion and the Arts*. New York: Oxford University Press, 1997.

Hockett, Charles F., and Robert Ascher. "The Human Revolution." *Current Anthropology* 5, (1964): 135–47.

Hovers, Erella, Shimon Ilani, Ofer Bar-Yosef, and Bernard Vandermeersch. "An Early Case of Color Symbolism: Ochre Use by Modern Humans in Qafzeh Cave." *Current Anthropology* 44, no. 4 (2003): 491–522. Accessed July 19, 2017. doi:10.1086/375869.

Humphrey, Nicholas. "The Social Function of Intellect." In *Growing Points in Ethology. Based on a Conference Sponsored by St. John's College and King's College, Cambridge*, edited by Paul P. G. Bateson and Robert A. Hinde, 303–17. Cambridge: Cambridge University Press, 1976.

Hunter, Patrick G., E. Glenn Schellenberg, and Ulrich Schimmack. "Mixed Affective Responses to Music with Conflicting Cues." *Cognition and Emotion* 22, no. 2 (2008): 327–52. Accessed July 26, 2017. doi:10.1080/02699930701438145.

Huron, David. "Is Music an Evolutionary Adaptation?"*Annals of the New York Academy of Sciences* 930, (2001): 43–61.

Hutcheson, Francis. *An Inquiry into the Original of Our Ideas of Beauty and Virtue*. In *Francis Hutcheson. Collected Works*, vol. 1, ed. Bernhard Fabian. Hildesheim: Olms, 1900.

Irons, William. "Religion as a Hard-to-Fake Sign of Commitment." In *Evolution and the Capacity for Commitment*, edited by Randolph M. Nesse, 292–309. New York: Russell Sage Foundation, 2001.

Istók, Eva, Elvira Brattico, Thomas Jacobsen, Kaisu Krohn, Mira Müller, and Mari Tervaniemi. "Aesthetic Responses to Music: A Questionnaire Study." *Musicae Scientiae* 13, no. 2 (2009): 183–206. Accessed July 26, 2017. doi:10.1177/102986490901300201.

Ivanhoe, Francis, and Erik Trinkaus. "On Cranial Deformation in Shanidar 1 and 5." *Current Anthropology* 24, no. 1 (1983): 127–28. Accessed August 10, 2017. doi:10.1086/202956.

Jacobson, Edith. "The 'Exceptions': An Elaboration of Freud's Character Study." *The Psychoanalytic Study of the Child* 14, no. 1 (1959): 135–54. Accessed September 26, 2017. doi:10.1080/00797308.1959.11822826.

Jacobsen, Thomas, Katharina Buchta, Michael Köhlerand, and Erich Schröger. "The Primacy of Beauty in Judging the Aesthetics of Objects." *Psychological Reports* 94 (2004): 1253–60. Accessed August 10, 2017. doi:10.2466/pr0.94.3c.1253-1260.

Jäger, Christoph and Georg Meggle, eds. *Kunst und Erkenntnis*. Paderborn: Mentis, 2005.

Jakobson, R. "Linguistics and Poetics." In *Style in Language*, edited by Thomas A. Sebeok, 350–77. Cambridge, MA: The MIT Press, 1960.

Jarvis, Erich D., Constance Scharff, Matthew R. Grossman, Joana A. Ramos, and Fernando Nottebohm. "For Whom The Bird Sings." *Neuron* 21, no. 4 (1998): 775–88. Accessed August 10, 2017. doi:10.1016/S0896-6273(00)80594-2.

Janik, Vincent M. and Peter J. Slater. "Vocal Learning in Mammals." *Advances in the Study of Behavior* 26 (1997): 59–99. Accessed August 10, 2017. doi:10.1016/S0065-3454(08)60377-0.

Jensvold, Mary L., and Deborah Fouts. "Imaginary Play in Chimpanzees (Pan Troglodytes)." *Human Evolution* 8, no. 3 (1993): 217–27. Accessed July 19, 2017. doi:10.1007/bf02436716

Johnson, Dan R. "Transportation into a Story Increases Empathy, Prosocial Behavior, and Perceptual Bias toward Fearful Expressions." *Personality and Individual Differences* 52, no. 2 (2012): 150–55.

Jones, Stephanie M., Joshua L. Brown, and J. Lawrence Aber. "Two-Year Impacts of a Universal School-Based Social-Emotional and Literacy Intervention: An Experiment in Translational Developmental Research." *Child Development* 82, no. 2 (2011): 533–54. Accessed August 10, 2017. doi:10.1111/j.1467-8624.2010.01560.x.

Jouary, Jean-Paul. *L'Art Paléolithique. Reflexions Philosophique*. Paris: Editions L'Harmattan, 2002.

Joyce, Robert. *The Esthetic Animal. Man, the Art-Created Art-Creator*. Hickswill, NY: Exposition Press, 1975.

Jürgens, Uwe. "Neural Pathways Underlying Vocal Control." *Neuroscience and Biobehavioral Reviews* 26, no. 2 (2002): 235–258. Accessed July 19, 2017. doi:10.1016/S0149-7634(01)00068-9.

Juslin, Patrik N., and Petri Laukka. "Communication of Emotions in Vocal Expression and Music Performance: Different Channels, Same Code?" *Psychological Bulletin* 129, no. 5 (2003): 770–814. Accessed July 19, 2017. doi:10.1037/0033-2909.129.5.770.

Justus, Timothy, and Jeffrey J. Hutsler. "Fundamental Issues in the Evolutionary Psychology of Music: Assessing Innateness and Domain Specificity." *Music Perception: An Interdisciplinary Journal* 23 (2005): 1–27.

Kant, Immanuel. *Critique of the Power of Judgment*. Translated by Paul Guyer and Eric Matthews. Cambridge: Cambridge University Press, 2000. First published 1790.

Kant, Immanuel. *Anthropology from a Pragmatic Point of View*. Translated by Mary G. Gregor. The Hague: Martinus Nijhoff, 1974.

Katz, Leonard D., ed. *Evolutionary Origins of Morality: Cross-Disciplinary Perspectives*. Exeter: Imprint Academic, 2000.

Kavolis, Vytautas. "The Value-Orientations Theory of Artistic Style." In *Art and Aesthetics in Primitive Societies: A Critical Anthology*, edited by Carol F. Jopling, 250–70. New York: E. P. Dutton & Co., 1971.

Kebeck, Günther and Henning Schroll. *Experimentelle Ästhetik*. Wien: Facultas, 2011.

Kidd, David C. and Emanuele Castano. "Reading Literary Fiction Improves Theory of Mind." *Science* 342, no. 6156 (2013): 377. Accessed July 27, 2017. doi:10.1126/science.1239918.

Kirschner, Sebastian, and Michael Tomasello. "Joint Music Making Promotes Prosocial Behavior in 4-Year-Old Children." *Evolution and Human Behavior* 31, no. 5 (2010): 354–64. Accessed July 27, 2017. doi:10.1016/j.evolhumbehav.2010.04.004.

Kittler, Friedrich A. *Discourse networks 1800/1900*. Translated by Michael Metteer. Stanford, CA: Stanford University Press, 1990.

Kittler, Ralf, Manfred Kayser, and Mark Stoneking. "Molecular Evolution of Pediculus Humanus and the Origin of Clothing." *Current Biology* 13, no. 16 (2003): 1414–17. Accessed July 19, 2017. doi:10.1016/S0960-9822(03)00507-4.

Kivy, Peter. "Charles Darwin on Music." *Journal of the American Musicological Society* 12, no. 1 (1959): 42–48. Accessed July 27, 2017. doi:10.2307/829516.

Kivy, Peter. "Herbert Spencer and a Musical Dispute." *Music Review* 23 (1964): 317–29.

Kleimann, Bernd, and Reinold Schmücker, eds. *Wozu Kunst? Die Frage nach ihrer Funktion*. Darmstadt: Wissenschaftliche Buchgesellschaft, 2001.

Knight, Chris. "Ritual Speech Coevolution: A Solution to the Problem of Deception." In *Approaches to the Evolution of Language*, edited by James R. Hurford, Michael Studdert-Kennedys and Chris Knight, 68–91. Cambridge: Cambridge University Press, 1998.

Knoop, Christine A., Valentin Wagner, Thomas Jacobsen, and Winfried Menninghaus. "Mapping the Aesthetic Space of Literature 'from below.'" *Poetics* 56 (2016): 35–49. Accessed July 28, 2017. doi:10.1016/j.poetic.2016.02.001.

Koelsch, Stefan, Thomas C. Gunter, D. Yves v. Cramon, Stefan Zysset, Gabriele Lohmann, and Angela D. Friederici. "Bach Speaks: A Cortical 'Language-Network' Serves the Processing of Music." *NeuroImage* 17, no. 2 (2002): 956–66. Accessed July 28, 2017. doi:10.1006/nimg.2002.1154.

Koelsch, Stefan, and Walter A. Siebel. "Towards a Neural Basis of Music Perception." *Trends in Cognitive Sciences* 9, no. 12 (2005): 578–84. Accessed August 8, 2017. doi:10.1016/j.tics.2005.10.001.

Koelsch, Stefan. "From Social Contact to Social Cohesion—The 7 Cs." *Music and Medicine* 5, no. 4 (2013): 204–9. Accessed August 8, 2017. doi:10.1177/1943862113508588.

Kölbl, Stefanie, and Nicholas Conard, eds. *Eiszeitschmuck. Status und Schönheit*, Museumsheft 6. Blaubeuren: Urgeschichtliches Museum, 2003.

Kölbl, Stefanie. "Ich, Wir und die Anderen. Kleidung und Schmuck als Statement." *In Eiszeit: Kunst und Kultur*, edited by Susanne Rau and Archäologisches Landesmuseum Baden-Württemberg, 167–75. Ostfildern: Thorbecke, 2009.

Kohn, Marek, and Steven J. Mithen. "Handaxes: Products of Sexual Selection?" *Antiquity* 73, no. 281 (1999): 518–26. Accessed August 08, 2017. doi:10.1017/S0003598X00065078.

Kroodsma, Donald E. "Learning and the Ontogeny of Sound Signals in Birds." In *Acoustic Communication in Birds: Song Learning and its Consequences*, edited by Donald E. Kroodsma, Edward H. Miller, and Henri Quellet, 1–23. New York: Academic Press, 1982.

Kroodsma, Donald E. *The Singing Life of Birds: The Art and Science of Listening to Birdsong*. New York: Houghton Mifflin Company, 2005.

Kraft, Georg. *Der Urmensch als Schöpfer: Die geistige Welt des Eiszeitmenschen*. Tübingen: Dr. M. Matthiesen & Co. KG. 1948.

Krebs, John R., and Richard Dawkins. "Animal Signals: Mind-Reading and Manipulation." In *Behavioural Ecology. An Evolutionary Approach*, edited by John R Krebs and Nicholas Davies, 380–402. Oxford: Blackwell Scientific, 1984.

Kreide-Damani, Ingrid. *Kunst-Ethnologie: Zum Verständnis fremder Kunst*. Köln: DuMont, 1992.

Kubovy, Michael. "On the Pleasures of the Mind." In *Well-Being. The Foundations of Hedonic Psychology*, edited by Daniel Kahnemann, Ed Diener, and Norbert Schwarz, 134–54. New York: Russell Sage Foundation, 1999.

Kummer, Hans, Lorraine Daston, Gerd Gigerenzer, and Joan B. Silk. "The Social Intelligence Hypothesis." In *Human by Nature: Between Biology and the Social Sciences*, edited by Peter Weingart, Peter Richerson, Sandra D. Mitchell, and Sabine Maasen, 157–79. Mahwah, NJ: Lawrence Erlbaum Associates, 1997.

Kumschick, Irina Rosa, Luna Beck, Michael Eid, Georg Witte, Gisela Klann-Delius, Winfried Menninghaus, Isabella Heuser, and Ruediger Steinlein. "Reading and Feeling: The Effects of a Literature-Based Intervention Designed to Increase Emotional Competence in Second and Third Graders." *Frontiers in Psychology* 5 (2014). Accessed August 8, 2017. doi:10.3389/fpsyg.2014.01448.

Lai, Cecilia S. L., Simon E. Fisher, Jane A. Hurst, Faraneh Vargha-Khadem, and Anthony P. Monaco. "A Forkhead-Domain Gene is Mutated in a Severe Speech and Language Disorder." *Nature* 413, no. 6855 (2001): 519–23. Accessed July 28, 2017. doi:10.1038/35097076.

Lande, Russell. "Sexual Dimorphism, Sexual Selection, and Adaption in Polygenic Characters." *Evolution* 34, no. 2 (1980): 292–305. Accessed August 8, 2017. doi:10.1111/j.1558-5646.1980.tb04817.x.

Lande, Russell. "Models of Speciation by Sexual Selection on Polygenic Traits." *Proceedings of the National Academy of the Sciences USA* 78, no. 6 (1981): 3721–25.

Lande, Russell. "Genetic Correlations between the Sexes in the Evolution of Sexual Dimorphism and Mating Preferences." In *Dahlem Workshop Reports - Sexual Selection: Testing the Alternatives*, edited by Jack W. Bradbury and Malte B. Andersson, 83–94. Chichester: John Wiley & Sons, 1987.

Langlois, Judith H., Lori A. Roggman, Rita J. Casey, Jean M. Ritter, Loretta A. Rieser-Danner, and Vivian Y. Jenkins. "Infant Preferences for Attractive Faces: Rudiments of a Stereotype?" *Developmental Psychology* 23, no. 3 (1987): 363–69. Accessed July 19, 2017. doi:10.1037//0012-1649.23.3.363.

Langlois, Judith H., Lori A. Roggman, and Lisa Musselman. "What Is Average and What Is Not Average About Attractive Faces?" *Psychological Science* 5, no. 4 (1994): 214–220. Accessed July 19, 2017. doi:10.1111/j.1467-9280.1994.tb00503.x.

Langmore, Naomi. E., Nicholas B. Davies, Ben J. Hatchwell, and Ian R. Hartley. "Female Song Attracts Males in the Alpine Accentor Prunella Collaris." *Proceedings: Biological Sciences* 263, no. 1367 (1996): 141–46.

Langmore, Naomi E. "Why Female Birds Sing." In *Signalling and Signal Design in Animal Communication*, edited by Yngve Espmark, Trond Amundsen, and Gunilla Rosenqvist, 317–27. Trondheim: Tapir Academic Press, 2000.

Leinonen, Lea, Ilkka Linnankoski, Maija-Liisa Laakso, and Reijo Aulanko. "Vocal Communication between Species: Man and Macaque." *Language & Communication* 11, no. 4 (1991): 241–62. Accessed August 18, 2017. doi:10.1016/0271-5309(91)90031-P.

Leinonen, L., M. L. Laakso, S. Carlson, and I. Linnankoski. "Shared Means and Meanings in Vocal Expression of Man and Macaque." *Logopedics Phoniatrics Vocology* 28, no. 2 (2003): 53–61. Accessed August 18, 2017. doi:10.1080/14015430310011754.

Leitão, Albertine, and Katharina Riebel. "Are Good Ornaments Bad Armaments? Male Chaffinch Perception of Songs with Varying Flourish Length." *Animal Behaviour* 66, no. 1 (2003): 161–67. Accessed August 18, 2017. doi:10.1006/anbe.2003.2167.

Leitch, Thomas M. *What Stories Are: Narrative Theory and Interpretation*. University Park: Pennsylvania State University Press. 1986.

Lenain, Thierry. "Ape-Painting and the Problem of the Origin of Art." *Human Evolution* 10, no. 3 (1995): 205–15. Accessed July 26, 2017. doi:10.1007/bf02438973.

Lethmate, Jürgen. "Tool-Using Skills of Orang-Utans." *Journal of Human Evolution* 11, no. 1 (1982): 49–64. Accessed July 26, 2017. doi:10.1016/S0047-2484(82)80031-6.

Lévi-Strauss, Claude. "The Science of the Concrete." In *Art and Aesthetics in Primitive Societies: A Critical Anthology*, edited by Carol F. Jopling, 225–49. New York: E. P. Dutton & Co., 1971.

Levinson, Jerrold, ed. *Ethics and Aesthetics: Essays at the Intersection*. Cambridge: Cambridge University Press, 1998.

Levinson, David, and Melvin Ember. *Encyclopedia of Cultural Anthropology*. Vol. 1. New York: Henry Holt and Co., 1996.

Levinson, Stephen C. "On the Human 'Interaction Engine.'" In *Roots of Human Sociality. Culture, Cognition and Interaction*, edited by Nicholas J. Enfield and Stephen C. Levinson, 39–69. New York: Berg, 2006.

Levitin, Daniel J., and Vidod Menon. "Musical Structure Is Processed in 'Language' Areas of the Brain: A Possible Role for Brodmann Area 47 in Temporal Coherence." *NeuroImage* 20, no. 4 (2003): 2142–52. Accessed July 26, 2017. doi:10.1016/j.neuroimage.2003.08.016.

Lewis-Williams, J. David, and Thomas A. Dowson. "The Signs of All Times: Entoptic Phenomena in Upper Paleolithic Art," *Current Anthropology* 29, no. 2 (1988): 201–217.

Lewis-Williams, David L. *The Mind in the Cave: Consciousness and the Origins of Art*. London: Thames & Hudson. 2002.

Lillard, Angeline. "Pretend Play as Twin Earth: A Social-Cognitive Analysis." *Developmental Review* 21, no. 4 (2001): 495–531.

Lim, Vanessa K., John L. Bradshaw, Michael E. R. Nicholls, Ian J. Kirk, Jeff P. Hamm, Michael Grossbach, and Eckart Altenmüller. "Aberrant Sensorimotor Integration in Musicians' Cramp Patients." *Journal of Psychophysiology* 17, no. 4 (2003): 195–202.

Linné, Sigvald. "Dental Decoration in Aboriginal America." *Ethnos* 5, nos., 1–2 (1940): 2–28. Accessed July 19, 2017. doi:10.1080/00141844.1940.9980567.

Lorblanchet, Michel. *Höhlenmalerei. Ein Handbuch.* Translated by Peter Nittmann. Sigmaringen: Thorbecke, 1997.

Lorenz, Konrad. "Die angeborenen Formen menschlicher Erfahrung." *Zeitschrift für Tierpsychologie* 5, no. 2 (1943): 235–409.

Low, Bobbi S. "Sexual Selection and Human Ornamentation." In *Evolutionary Biology and Human Social Behavior: An Anthropological Perspective*, edited by Napoleon A. Chagnon and William Irons, 462–87. North Scituate, MA: Duxbury Press, 1979.

Löw, Reinhard. *Philosophie des Lebendigen: der Begriffe des Organischen bei Kant, sein Grund und seine Aktualität.* Frankfurt am Main: Suhrkamp. 1980.

Lyn, Heidi, Patricia Greenfield, and Sue Savage-Rumbaugh. "The Development of Representational Play in Chimpanzees and Bonobos: Evolutionary Implications, Pretense, and the Role of Interspecies Communication." *Cognitive Development* 21, no. 3 (2006): 199–213. Accessed July 19, 2017. doi:10.1016/j.cogdev.2006.03.005.

Mace, Ruth, Claire J. Holden, and Stephen Shennan, eds. *The Evolution of Cultural Diversity: A Phylogenetic Approach.* Walnut Creek, CA: Left Coast Press, 2005.

Maess, Burkhard, Stefan Koelsch, Thomas C. Gunter, and Angela D. Friederici. "Musical Syntax Is Processed in Broca's Area: An MEG Study." *Nature Neuroscience* 4, no. 5 (2001): 540–45. Accessed July 19, 2017. doi:10.1038/87502.

Mar, Raymond A., Keith Oatley, Jacob Hirsh, Jennifer dela Paz, and Jordan B. Peterson. "Bookworms versus Nerds: Exposure to Fiction versus Non-Fiction, Divergent Associations with Social Ability, and the Simulation of Fictional Social Worlds." *Journal of Research in Personality* 40, no. 5 (2006): 694–712. Accessed August 10, 2017. doi:10.1016/j.jrp.2005.08.002.

Marler, Peter R., and Hans Slabbekoorn, eds. *Nature's Music: The Science of Birdsong.* New York: Elsevier Academic Press, 2004.

Marler, Peter. "Origins of Music and Speech: Insights from Animals." In *The Origins of Music*, edited by Nils L. Walin, Björn Merker and Steven Brown, 31-48. Cambridge, MA: MIT Press, 2001.

Martindale, Colin, and Kathleen Moore. "Priming, Prototypicality, and Preference." *Journal of Experimental Psychology: Human Perception and Performance* 14, no. 4 (1988): 661–70. Accessed August 10, 2017. doi:10.1037//0096-1523.14.4.661.

Martindale, Colin, Kathleen Moore, and Jonathan Borkum. "Aesthetic Preference: Anomalous Findings for Berlyne's Psychobiological Theory." *The American Journal of Psychology* 103, no. 1 (1990): 53–80.

Martinelli, Dario. "Liars, Players, and Artists: A Zoosemiotic Approach to Aesthetics." *Semiotica* 150 (2004): 77–118.

Mathy, Dietrich. *Poesie und Chaos. Zur anarchistischen Komponente der frühromantischen Ästhetik*. München: Weixler, 1984.

McDermott, Josh, and Marc Hauser. "The Origins of Music: Innateness, Uniqueness, and Evolution." Music Perception: An Interdisciplinary Journal 23, no. 1 (2005): 29–59.

McNeill, William. *Keeping Together in Time: Dance and Drill in Human History*. Cambridge, MA: Harvard University Press, 1995.

McPherron, Shannon P., Zeresenay Alemseged, Curtis W. Marean, Jonathan G. Wynn, Denne Reed, Denis Geraads, Rene Bobe, and Hamdallah A. Bearat. "Evidence for Stone-Tool-Assisted Consumption of Animal Tissues before 3.39 Million Years Ago at Dikika, Ethiopia." *Nature* 466, no. 7308 (2010): 857–60. Accessed July 26, 2017. doi:10.1038/nature09248.

Mellars, Paul. *The Neandertal Legacy*: An Archaeological Perspective from Western Europe. Princeton, NJ: Princeton University Press, 1996.

Meltzoff, Andrew N. "Towards a Developmental Cognitive Science: The Implications of Cross-Modal Matching and Imitation for the Development of Representation and Memory in Infancy." *Annals of the New York Academy of Sciences* 608, no. 1 (1990): 1–37. Accessed August 17, 2017. doi:10.1111/j.1749-6632.1990.tb48889.x.

Menninghaus, Winfried. "Mitologia do Caos no Romantismo e na Modernidade." *Estudos Avançados* 10, no. 27 (1996): 127–38.

Menninghaus, Winfried. *In Praise of Nonsense: Kant and Bluebeard*. Edited by Werner Hamacher and David E. Wellbery. Translated by Henry Pickford. Meridian: Crossing Aesthetics. Stanford, CA: Stanford University Press. 1999.

Menninghaus, Winfried. *Das Versprechen der Schönheit*. Frankfurt am Main: Suhrkamp, 2003.

Menninghaus, Winfried. *Kunst als "Beförderung des Lebens": Perspektiven transzendentaler und evolutionärer Ästhetik, Themen - Carl Friedrich von Siemens Stiftung*. Vol. 89. München: Carl Friedrich von Siemens Stiftung. 2008

Menninghaus, Winfried. "'Ein Gefühl der Beförderung des Lebens'. Kants Reformulierung des Topos lebhafter Vorstellung." In *Vita aesthetica. Szenarien ästhetischer Lebendigkeit*, edited by Armen Avanessian, Heinz Lunzer and Jan Völker, 77–94. Zürich: Diaphanes 2009.

Menninghaus, Winfried. "Biologie nach der Mode. Charles Darwins Ornament-Ästhetik." In *Evolution und Kultur des Menschen*, edited by Ernst P. Fischer and Klaus Wiegandt, 220–43. Frankfurt am Main: Fischer Taschenbuchverlag, 2010.

Menninghaus, Winfried, Isabel C. Bohrn, Ulrike Altmann, Oliver Lubrich, and Arthur M. Jacobs. "Sounds Funny? Humor Effects of Phonological and Prosodic Figures of Speech." *Psychology of Aesthetics Creativity and the Arts* 8, no. 1 (2014): 71–76. Accessed December 17, 2014. doi:10.1037/a0035309.

Menninghaus, Winfried, Isabel C. Bohrn, Christine A. Knoop, Sonja A. Kotz, Wolff Schlotz, and Arthur M. Jacobs. "Rhetorical Features Facilitate Prosodic Processing While Handicapping Ease of Semantic Comprehension." *Cognition* 143 (2015): 48–60.

Menninghaus, Winfried, Valentin Wagner, Eugen Wassiliwizky, Thomas Jacobsen, and Christine A. Knoop. "The Emotional and Aesthetic Powers of Parallelistic Diction." Poetics 63 (2017): 47–59. Accessed September 4, 2017. doi:10.1016/j.poetic.2016.12.001

Menninghaus, Winfried, Valentin Wagner, Julian Hanich, Eugen Wassiliwizky, Thomas Jacobsen, and Stefan Koelsch. "The Distancing–Embracing Model of the Enjoyment of Negative Emotions in Art Reception." Behavioral and Brain Sciences (2017): 1–58. Accessed August 10, 2017. doi:10.1017/S0140525X17000309.

Merker, Björn. "Synchronous Chorussing and Human Origins." In *The Origins of Music*, edited by Nils L. Wallin, Björn Merker and Steven Brown, 315–28. Cambridge, MA: The MIT Press, 2000.

Miall, David S., and Ellen Dissanayake. "The Poetics of Babytalk." *Human Nature* 14, no. 4 (2003): 337–64. Accessed December 17, 2014. doi:10.1007/s12110-003-1010-4.

Miller, Geoffrey. "Evolution of Human Music through Sexual Selection." In *The Origins of Music*, edited by Nils L. Wallin, Björn Merker, and Steven Brown, 329–60. Cambridge, MA: The MIT Press, 2000.

Miller, Geoffrey. *The Mating Mind: How Sexual Choice Shaped the Evolution of Human Nature*. New York: Anchor Books, 2000.

Mitani, John C., Toshikazu Hasegawa, Julie Gros-Louis, Peter Marler, and Richard Byrne. "Dialects in Wild Chimpanzees?" *American Journal of Primatology* 27, no. 4 (1992): 233–43. Accessed August 10, 2017. doi:10.1002/ajp.1350270402.

Mitchell, Robert W., ed. *Pretending and Imagination in Animals and Children*. Cambridge: Cambridge University Press, 2002.

Mithen, Steven. *The Prehistory of the Mind. A Search for the Origins of Art, Religion, and Science*. London: Phoenix, 1998.

Mithen, Steven. *The Singing Neanderthals. The Origins of Music, Language, Mind, and Body*. Cambridge, MA: Harvard University Press, 2007.

Molino, Jean. "Toward an Evolutionary Theory of Music and Language." In *The Origins of Music*, edited by Nils L. Wallin, Björn Merker, and Steven Brown, 165–76. Cambridge, MA: The MIT Press, 2000.

Montagna, William. "The Evolution of Human Skin(?)." *Journal of Human Evolution* 14, no. 1 (1985): 3–22. Accessed September 7, 2017. doi:10.1016/S0047-2484(85)80090-7.

Moritz, Karl Philipp. "Die Signatur des Schönen. Inwiefern Kunstwerke beschrieben werden können." In *Werke*, edited by Horst Günther, 579–88. Vol. 2. Frankfurt am Main: Insel Verlag, 1981.

Morreall, Jon. "Enjoying Negative Emotions in Fictions." *Philosophy and Literature* 9, no. 1 (1985): 95–103.

Moretti, Franco. *Graphs, Maps, Trees: Abstract Models for a Literary History*. London: Verso, 2007.

Morris, Desmond. *The Biology of Art. A Study of the Picture-Making Behaviour of the Great Apes and Its Relationship to Human Art*. London: Methuen, 1962.

Morris, Desmond. *The Naked Ape: A Zoologist's Study of the Human Animal.* London: Jonathan Cape., 1967.

Mothersbaugh, David L., Bruce A. Huhmann, and George R. Franke. "Combinatory and Separative Effects of Rhetorical Figures on Consumers' Effort and Focus in Ad Processing." *Journal of Consumer Research* 28, no. 4 (2002): 589–602. Accessed August 10, 2017. doi:10.1086/338211.

Mülder-Bach, Inka. *Im Zeichen Pygmalions. Das Modell der Statue und die Entdeckung der "Darstellung" im 18. Jahrhundert.* München: Fink, 1998.

Muensterberger, Warner. "Some Elements of Artistic Creativity Among Primitive Peoples." In *Art and Aesthetics in Primitive Societies: A Critical Anthology*, edited by Carol F. Jopling, 3–10. New York: E. P. Dutton & Co., 1971.

Müller-Sievers, Helmut, *Self-Generation. Biology, Philosophy, and Literature around 1800.* Stanford, CA: Stanford University Press, 1997.

Munte, Thomas F., Eckart Altenmuller, and Lutz Jancke. "The Musician's Brain as a Model of Neuroplasticity." *Nature Reviews Neuroscience* 3, no. 6 (2002): 473–78.

Nettl, Bruno. *The Study of Ethnomusicology: Twenty-Nine Issues and Concepts.* Urbana: University of Illinois Press, 1983.

Neuhaus, Christiane, Thomas R. Knösche, and Angela D. Friederici. "Effects of Musical Expertise and Boundary Markers on Phrase Perception in Music." *Journal of Cognitive Neuroscience* 18, no. 3 (2006): 472–93. Accessed August 18, 2017. doi:10.1162/jocn.2006.18.3.472

Neumann, Eckhardt. *Funktionshistorische Anthropologie der ästhetischen Produktivität.* Berlin: Reimer, 1996.

Nishida, Toshisada. "Sexual Behavior of Adult Male Chimpanzees of the Mahale Mountains National Park, Tanzania." *Primates* 38, no. 4 (1997): 379–98. Accessed August 18, 2017. doi:10.1007/bf02381879.

Nishida, Toshisada. "The Leaf-Clipping Display: A Newly-Discovered Expressive Gesture in Wild Chimpanzees." *Journal of Human Evolution* 9, no. 2 (1980): 117–28. Accessed August 8, 2017. doi:10.1016/0047-2484(80)90068-8.

Nottebohm, Fernando. "From Bird Song to Neurogenesis." *Scientific American* 260, no. 2 (1989): 74–79.

Nussbaum, Martha C. *Love's Knowledge. Essays on Philosophy and Literature.* Oxford: Oxford University Press, 1990.

Nussbaum, Martha C. *Poetic Justice. The Literary Imagination and Public Life.* Boston: Beacon Press, 1997.

Obermeier, Christian, Sonja A. Kotz, Sarah Jessen, Tim Raettig, Martin von Koppenfels, and Winfried Menninghaus. "Aesthetic Appreciation of Poetry Correlates with Ease of Processing in Event-Related Potentials." *Cognitive, Affective, & Behavioral Neuroscience* 16, no. 2 (2016): 362–73. Accessed August 8, 2017. doi:10.3758/s13415-015-0396-x.

Odling-Smee, F. John, Kevin N. Laland, and Marcus W. Feldman. *Niche Construction. The Neglected Process in Evolution.* Princeton, NJ: Princeton University Press, 2003.

Oesterle, Günter. "Vorbegriffe zu einer Theorie der Ornamente. Kontroverse Formprobleme zwischen Aufklärung, Klassizismus und Romantik am Beispiel der Arabeske." In *Ideal und Wirklichkeit der bildenden Kunst im späten 18. Jahrhundert*, edited by Herbert Beck, Peter C. Bol, and Eva Mack-Gérard, 119–40. Berlin: Gebr. Mann, 1984.

Orians, Gordon H., and Judith H. Heerwagen. "Evolved Responses to Landscapes." In *The Adapted Mind: Evolutionary Psychology and the Generation of Culture*, edited by Jerome H. Barkow, Leda Cosmides and John Tooby, 555–80. New York: Oxford University Press, 1995.

O'Sullivan, Noreen, Philip Davis, Josie Billington, Victorina Gonzalez-Diaz, and Rhiannon Corcoran. "'Shall I compare thee': The Neural Basis of Literary Awareness, and Its Benefits to Cognition." *Cortex* 73 (2015): 144–57. Accessed July 27, 2017. doi:10.1016/j.cortex.2015.08.014.

Owens, Ian P. F., Terry Burke, and B. A. Thompson Desmond. "Extraordinary Sex Roles in the Eurasian Dotterel: Female Mating Arenas, Female-Female Competition, and Female Mate Choice." *The American Naturalist* 144, no. 1 (1994): 76–100.

Pagel, Mark, and Walter Bodmer. "A Naked Ape Would Have Fewer Parasites." *Proceedings of the Royal Society of London B: Biological Sciences* 270 (2003): 117–19.

Pahnke, Walter M. "Drogen und Mystik." In *Religion und die Droge: ein Symposion über religiöse Erfahrungen unter Einfluss von Halluzinogenen*, edited by Manfred Josuttis and Hanscarl Leuner, 54–76. Stuttgart: Kohlhammer, 1972.

Panksepp, Jaak. "The Ontogeny of Play in Rats." *Developmental Psychobiology* 14, no. 4 (1981): 327–32. Accessed July 27, 2017. doi:10.1002/dev.420140405.

Panksepp, Jaak, Steve Siviy, and Larry Normansell. "The Psychobiology of Play: Theoretical and Methodological Perspectives." *Neuroscience and Behavioral Reviews* 8, no. 4 (1984): 465–92. Accessed July 27, 2017. doi:10.1016/0149-7634(84)90005-8.

Patel, Aniruddh D., Edward Gibson, Jennifer Ratner, Mireille Besson, and Phillip J. Holcomb. "Processing Syntactic Relations in Language and Music: An Event-Related Potential Study." *Journal of Cognitive Neuroscience* 10, no. 6 (1998): 717–33. Accessed July 27, 2017. doi:10.1162/089892998563121.

Patel, Aniruddh D. "Language, Music, Syntax and the Brain." *Nature Neuroscience* 6, no. 7 (2003): 674–81.

Patel, Aniruddh D. *Music, Language, and the Brain*. New York: Oxford University Press, 2008.

Patel, Aniruddh D. "Music, Biological Evolution, and the Brain." In *Emerging Disciplines: Shaping New Fields of Scholarly Inquiry in and Beyond the Humanities*, edited by Melissa Bailar, 41–64. Houston: Connexions - Rice University Press, 2010.

Parker, Scott, Jesse Bascom, Brian Rabinovitz, and Debra Zellner. "Positive and Negative Hedonic Contrast with Musical Stimuli." *Psychology of Aesthetics, Creativity, and the Arts* 2, no. 3 (2008): 171–74. Accessed July 11, 2017. doi:10.1037, 1931-3896.2.3.171.

Payne, Robert B., Laura L. Payne, and Susan M. Doehlert. "Biological and Cultural Success of Song Memes in Indigo Buntings." *Ecology* 69, no. 1 (1988): 104–17. Accessed July 11, 2017. doi:10.2307/1943165.

Peer, Willie van. "The Measurement of Metre. Its Cognitive and Affective Functions." *Poetics* 19, no. 3 (1990): 259–75. Accessed July 18, 2017. doi:10.1016/0304-422X(90)90023-X.

Peirce, Charles S., and Victoria Welby. *Semiotics and Significs: The Correspondence between Charles S. Peirce and Lady Victoria Welby*. Edited by Charles S. Hardwick and James Cook. Bloomington: Indiana University Press. 1977.

Perrett, David I., Keith May, and Sakiko Yoshikawa. "Facial Shape and Judgments of Female Attractiveness." *Nature* 368, no. 6468 (1994): 239–42.

Petrie, Marion. "Mating Decisions by Female Common Moorhens (Gallinula Choropus)." In *Acta XIX Congressus Internationalis Orhitologici*, edited by Henri Ouellet, 947–55. Ottawa: University of Ontario Press, 1986.

Pfeiffer, John E. *The Creative Explosion. An Inquiry into the Origins of Art and Religion*. New York: Harper & Row Publishers, 1982.

Pika, Simone, Katja Liebal, and Michael Tomasello. "Gestural Communication in Young Gorillas (Gorilla gorilla): Gestural Repertoire, Learning, and Use." *American Journal of Primatology* 60, no. 3 (2003): 95–111.

Pinker, Steven. *How the Mind Works*. London: Penguin, 1999.

Pinker, Steven. "Towards a Consilient Study of Literature." *Philosophy and Literature* 31, no. 1 (2007): 162–78.

Plutarch. *De gloria Atheniensum* 5, Moralia 348 C. Edited by Frank Cole Babbitt. Cambridge, MA: Harvard University Press.

Power, Camilla. "Old Wives' Tale: The Gossip Hypothesis and the Reliability of Cheap Signals." In *Approaches to the Evolution of Language*, edited by James R Hurford, Michael Studdert-Kennedy, and Chris Knight, 111–29. Cambridge: Cambridge University Press, 1998.

Power, Camilla. "Beauty Magic': The Origins of Art." In *The Evolution of Culture, An Interdisciplinary View*, edited by Robin Dunbar, Chris Knight, and Camilla Power, 92–112. Edinburgh: Edinburgh Press, 1999.

Pruetz, Jill D., Paco Bertolani, Kelly Boyer Ontl, Stacy Lindshield, Mack Shelley, and Erin G. Wessling. "New Evidence on the Tool-Assisted Hunting Exhibited by Chimpanzees (Pan troglodytes verus) in a Savannah Habitat at Fongoli, Sénégal." *Royal Society Open Science* 2, no. 4 (2015). Accessed October 4, 2017. doi:10.1098/rsos.140507.

Prum, Richard O. "Coevolutionary Aesthetics in Human and Biotic Artworlds." *Biology and Philosophy* 28, no. 5 (2013): 811–32. Accessed March 6, 2019. doi: 10.1007/s10539-013-9389-8.

Prum, Richard O. *The Evolution of Beauty: How Darwin's Forgotten Theory of Mate Choice Shapes the Natural World—and Us*. New York: Doubleday. 2017.

Queiroz do Amaral, Lia. "Loss of Body Hair, Bipedality and Thermoregulation. Comments on Recent Papers in the Journal of Human Evolution." *Journal of Human Evolution* 30, no. 4 (1996): 357–66. Accessed July 18, 2017. doi:10.1006/jhev.1996.0029.

Ramachandran, Vilayanur, S. and William Hirstein. "The Science of Art: A Neurological Theory of Aesthetic Experience." *Journal of Consciousness Studies* 6, nos. 6–7 (1999): 15–51.

Ramachandran, Vilayanur. *The Emerging Mind, Reith Lectures, 2003.* London: Profile, 2003.

Rantala, Markus J. "Human Nakedness: Adaptation against Ectoparasites?" *International Journal for Parasitology* 29, no. 12 (1999): 1987–89. Accessed July 18, 2017. doi:10.1016/S0020-7519(99)00133-2.

Rantala, Markus J. "Evolution of Nakedness in Homo Sapiens." *Journal of Zoology* 273, no. 1 (2007): 1–7.

Rau, Susanne, and Archäologisches Landesmuseum Baden-Württemberg, eds. *Eiszeit: Kunst und Kultur.* Ostfildern: Thorbecke, 2009.

Raphael, Max, ed. *Prähistorische Höhlenmalerei Aufsätze, Briefe.* Köln: Bruckner & Thünker, 1993.

Reber, Rolf, Piotr Winkielman, and Norbert Schwarz. "Effects of Perceptual Fluency on Affective Judgments." *Psychological Science* 9, no. 1 (1998): 45–48. Accessed August 10, 2017. doi:10.1111/1467-9280.00008.

Reber, Rolf, Norbert Schwarz, and Piotr Winkielman. "Processing Fluency and Aesthetic Pleasure: Is Beauty in the Perceiver's Processing Experience?" *Personality and Social Psychology Review* 8, no. 4 (2004): 364–82. Accessed July 28, 2017. doi:10.1207/s15327957pspr0804_3.

Rebora, Alfredo. "Lucy's Pelt: When We Became Hairless and How We Managed to Survive." *International Journal of Dermatology* 49, no. 1 (2010): 17–20. Accessed July 28, 2017. doi:10.1111/j.1365-4632.2009.04266.x.

Reichholf, Josef H. *Der Ursprung der Schönheit: Darwins größtes Dilemma.* München: C. H. Beck. 2011.

Redies, Christoph, Anselm Brachmann, and Johan Wagemans. "High Entropy of Edge Orientations Characterizes Visual Artworks from Diverse Cultural Backgrounds." *Vision Research* 133 (2017): 130–44.

Reimbold, Ernst T. *Der Pfau: Mythologie und Symbolik.* München: Callwey, 1983.

Rensch, Bernhard. "Ästhetische Faktoren bei Farb- und Formbevorzugungen von Affen." *Zeitschrift für Tierpsychologie* 14, (1957): 71–99. Accessed July 28, 2017. doi:10.1111/j.1439-0310.1957.tb00526.x.

Rensch, Bernhard. "Malversuche mit Affen." *Zeitschrift für Tierpsychologie* 18 (1961): 347–64. Accessed July 25, 2017. doi:10.1111/j.1439-0310.1961.tb00425.x.

Rensch, Bernhard. "Ästhetische Grundprinzipien bei Mensch und Tier." In *Kreatur Mensch. Moderne Wissenschaft auf der Suche nach dem Humanum*, edited by Günter Altner, 265–88. München: Deutscher Taschenbuchverlag, 1973.

Rensch, Bernhard. "Play and Art in Apes and Monkeys." *Symposia of the Fourth International Congress of Primatology* 1 (1973): 102–23.

Rensch, Bernhard. "Über ästhetische Faktoren im Erleben höherer Tiere." In *Evolution II. Ein Querschnitt der Forschung*, edited by Hoimer von Ditfurth, 229–46. Hamburg: Hoffmann und Campe, 1978.

Rentschler, Ingo, Barbara Herzberger, and David Epstein. *Beauty and the Brain. Biological Aspects of Aesthetics.* Basel: Birkhäuser, 1988.

Richman, Bruce. "How Music Fixed 'Nonsense' into Significant Formulas: On Rhythm, Repetition, and Meaning." In *The Origins of Music*, edited by Nils L. Walin, Björn Merker and Steven Brown, 301-14. Cambridge, MA: MIT Press, 2001.

Richman, Bruce. "Rhythm and Melody in Gelada Vocal Exchanges." *Primates* 28, no. 2 (1987): 199–223. Accessed August 22, 2017. doi:10.1007/bf02382570.

Richter, Klaus. *Die Herkunft des Schönen: Grundzüge der evolutionären Ästhetik*. Mainz: Philipp von Zabern, 1999.

Riebel, Katharina. "The 'Mute' Sex Revisited: Vocal Production and Perception Learning in Female Songbirds." *Advances in the Study of Behavior* 33 (2003): 49–86. Accessed August 22, 2017. doi:10.1016/S0065-3454(03)33002-5.

Rimé, Bernard. "Emotion Elicits the Social Sharing of Emotion: Theory and Empirical Review." *Emotion Review* 1, no. 1 (2009): 60–85. Accessed July 27, 2017. doi:10.1177/1754073908097189.

Ritchison, Gary. "The Singing Behavior of Female Northern Cardinals." *The Condor* 88, no. 2 (1986): 156–59. Accessed July 28, 2017. doi:10.2307/1368910.

Ritter, Johann Wilhelm, ed. *Fragmente aus dem Nachlaß eines jungen Physikers. Ein Taschenbuch für Freunde der Natur*. Vol. 1. Heidelberg: Mohr und Zimmer, 1810.

Robinson, Jenefer. *Deeper than Reason: Emotion and its Role in Literature, Music, and Art*. Oxford: Oxford University Press, 2007.

Roederer, Juan G. "The Search for a Survival Value for Music." *Music Perception: An Interdisciplinary Journal* 1, no. 3 (1984): 350–56. Accessed July 12, 2017. doi:10.2307/40285265.

Rogers, Spencer L. *Artificial Deformation of the Head: New World Examples of Ethnic Mutilation and Notes on Its Consequences*. San Diego, CA: Museum of Man, 1975.

Romero Molina, Javier. "Dental Mutilations, Trephination, and Cranial Deformation." In *Handbook of Middle American Indians*, edited by Robert Wauchope, 50-67. Vol. 9. Austin: University of Texas Press, 1970.

Rosenkranz, Karin, Eckart Altenmüller, Sabine Siggelkow, and Reinhard Dengler. "Alteration of Sensorimotor Integration in Musician's Cramp: Impaired Focussing of Proprioception." *Electronic Clinic Neurophysiology* 111, no. 11 (2000), 2036–41.

Rothenberg, David. *Why Birds Sing. A Journey into the Mystery of Bird Song*. New York: Basic Books, 2005.

Ryan, Michael J. "Sexual Selection, Sensory Systems, and Sensory Exploitation." *Oxford Survey of Evolutionary Biology* 7 (1990): 157–95.

Sahakian, William. *History and Systems of Psychology*. New York: Wiley & Sons 1975.

Samur, Dalya, Mattie Tops, and Sander L. Koole. "Does a Single Session of Reading Literary Fiction Prime Enhanced Mentalising Performance? Four Replication Experiments of Kidd and Castano (2013)." *Cognition and Emotion* (2017): 1–15. Accessed July 12, 2017. doi:10.1080/02699931.2017.1279591.

Sander, David, Didier Grandjean, and Klaus R. Scherer. "A Systems Approach to Appraisal Mechanisms in Emotion." *Neural Networks* 18, no. 4 (2005): 317–52.

Saranathan, Vinodkumar, Deborah Hamilton, George V. N. Powell, Donald E. Kroodsma, and Richard O. Prum. "Genetic Evidence Supports Song Learning in the Three-Wattled Bellbird Procnias Tricarunculata (Cotingidae)." *Molecular Ecology* 16, no. 17 (2007): 3689–702. Accessed July 28, 2017. doi:10.1111/j.1365-294X.2007.03415.x.

Sauvet, Susanne, Georges Sauvet, and André Wlodarczyk. "Essai de Sémiologie Préhistorique." *Bulletin de la Société Préhistorique Française* 74 (1977): 545–58.

Schellenberg, E. Glenn. "Music Lessons Enhance IQ." *Psychological Science* 15, no. 8 (2004): 511–14. Accessed August 10, 2017. doi:10.1111/j.0956-7976.2004.00711.x.

Schellenberg, E. Glenn. "Exposure to Music: The Truth about the Consequences." In *The Child as Musician: A Handbook of Musical Development*, edited by Gary E. McPherson, 111–34. Oxford: Oxford University Press, 2006.

Schellenberg, E. Glenn, Takayuki Nakata, Patrick G. Hunter, and Sachiko Tamoto. "Exposure to Music and Cognitive Performance: Tests of Children and Adults." *Psychology of Music* 35, no. 1 (2007): 5–19. Accessed August 10, 2017. doi:10.1177/0305735607068885.

Schellenberg, E. Glenn, Isabelle Peretz, and Sandrine Vieillard. "Liking for Happy- and Sad-Sounding Music: Effects of Exposure." Cognition *and Emotion* 22, no. 2 (2008): 218–37. Accessed July 28, 2017. doi:10.1080/02699930701350753.

Scherer, Klaus R. "Emotion Serves to Decouple Stimulus and Response." In *The Nature of Emotion. Fundamental Questions*, edited by Paul Ekman and Richard J. Davidson, 127–30. New York: Oxford University Press, 1994.

Scherer, Klaus R. "What Are Emotions? And How Can They Be Measured?" *Social Science Information* 44, no. 4 (2005): 695–729. Accessed July 28, 2017. doi:10.1177/0539018405058216.

Schlaug, Gottfried. "The Brain of Musicians." *Annals of the New York Academy of Sciences* 930, no. 1 (2001): 281–99.

Schmandt-Besserat, Denise. "Vom Ursprung der Schrift." *Spektrum der Wissenschaft* 12 (1978): 4–13.

Schmücker, Reinold. "Funktionen der Kunst." In *Wozu Kunst? Die Frage nach ihrer Funktion*, edited by Bernd Kleimann and Reinold Schmücker, 13–33. Darmstadt: Wissenschaftliche Buchgesellschaft, 2001.

Schögler, Ben, and Colwyn Trevarthen. "To Sing and Dance Together. From Infants to Jazz." In *On Being Moved. From Mirror Neurons to Empathy*, edited by Stein Bråten, 281–302. Philadelphia: John Benjamins Publishing Company, 2007.

Schrenk, Friedemann. "Vom aufrechten Gang zur Kunst. Die Entwicklung und Ausbreitung des Menschen." In *Eiszeit: Kunst und Kultur*, edited by Susanne Rau and Archäologisches Landesmuseum Baden-Württemberg, 52–59. Ostfildern: Thorbecke, 2009.

Schrott, Raoul, and Arthur Jacobs. *Gehirn und Gedicht. Wie wir unsere Wirklichkeiten konstruieren*. München: Hanser, 2011.

Schwabe, Julius. "Zur Symbolik des Urmenschen." In *Symbolon. Jahrbuch für Symbolforschung*, edited by Julius Schwabe, 190–242. Vol. 7. Basel: Schwabe & Co. Verlag, 1971.

Scruton, Roger. *The Aesthetics of Music*. Oxford: Clarendon Press, 2009.

Searcy, William A., and Stephen Nowicki. *The Evolution of Animal Communication: Reliability and Deception in Signaling Systems*. Princeton, NJ: Princeton University Press, 2005.

Selander, Robert K. "Sexual Selection and Dimorphism in Birds." In *Sexual Selection and the Descent of Man 1871–1971*, edited by Bernard Campbell, 180–229. London: Heineman, 1971.

Seligmann, Martin E. *Helplessness: On Depression, Development, and Death*. San Francisco: W. H. Freeman and Company, 1975.

Shklovsky, Viktor. "Art as Device." In *Viktor Shklovsky: A Reader*, edited and translated by Alexandra Berlina, 73–96. New York: Bloomsbury Academic, 2017.

Simmel, Georg. "Fashion." *The American Journal of Sociology* 62, no. 6 (1957), 541–58.

Skamel, Uta. "Beauty and Sex Appeal: Sexual Selection of Aesthetic Preferences." In *Evolutionary Aesthetics*, edited by Eckart Voland and Karl Grammer, 173–200. Heidelberg: Springer, 2003.

Slater, Peter J. B. "Birdsong Repertoires: Their Origins and Use." In *The Origins of Music*, edited by Nils L. Walin, Björn Merker and Steven Brown, 49-63. Cambridge, MA: MIT Press, 2001.

Smith, Greg M. "Local Emotions, Global Moods, and Film Structure." In *Passionate Views. Film, Cognition, and Emotion*, edited by Carl Plantinga and Greg M. Smith, 103–26. Baltimore, MD: Johns Hopkins University Press, 1999.

Smith, John Maynard. "Sexual Selection and the Handicap Principle." *Journal of Theoretical Biology* 57, no. 1 (1976): 239–42. Accessed July 27, 2017. doi:10.1016/S0022-5193(76)80016-1.

Smithrim, Katherine, and Rena Upitis. "Learning Through the Arts: Lessons of Engagement." *Canadian Journal of Education* 28, (2005): 109–27.

Snowdon, Charles T., and A. Margaret Elowson. "'Babbling' in Pygmy Marmosets: Development after Infancy." *Behaviour* 138, no. 10 (2001): 1235–48.

Soler, Charles, Manuel Núñez, Ricardo Gutiérrez, Javier Núñez, Pascual Medina, Maria Sancho, Juan Álvarez, and Nunez Álvarez. "Facial Attractiveness in Men Provides Clues to Semen Quality." *Evolution and Human Behavior* 24, no. 3 (2003): 199–207. Accessed July 28, 2017. doi:10.1016/S1090-5138(03)00013-8.

Sommer, Volker. "Die Vergangenheit einer Illusion. Religion aus evolutionsbiologischer Sicht." In *Evolution und Anpassung: Warum die Vergangenheit die Gegenwart erklärt: Christian Vogel zum 60. Geburtstag*, edited by Eckart Voland, 229–48. Stuttgart: S. Hirzel Verlag, 1993.

Sopčák, Paul, Massimo Salgaro, and J. Berenike Herrmann, eds. *Transdisciplinary Approaches to Literature and Empathy*. Vol. 6, *Scientific Study of Literature*. Amsterdam: John Benjamins Publishing Company, 2016.

Sosis, Richard. "Why Aren't We All Hutterites? Costly Signaling Theory and Religious Behavior." *Human Nature* 14, no. 2 (2003): 91–127. Accessed July 27, 2017. doi:10.1007/s12110-003-1000-6.

Spencer, Herbert. "The Origin and Function of Music." *Essays: Scientific, Political and Speculative*, 359-384. Vol. 1. London: Longman, Brown, Green, Longmans & Roberts, 1858.

Spencer, Herbert. *Ceremonial Institutions. The Principles of Sociology*. Vol. 2, part 4. London: Williams & Norgate, 1879.

Spencer, Herbert. "The Origin of Music." *Mind. A Quarterly Review of Psychology and Philosophy* 60 (1890): 449–68.

Sperber, Dan, and Lawrence A. Hirschfeld. "The Cognitive Foundations of Cultural Stability and Diversity." *Trends in Cognitive Sciences* 8, no. 1 (2004): 40–46. Accessed August 10, 2017. doi:10.1016/j.tics.2003.11.002.

Standley, Jayne M. "The Effect of Music and Multimodal Stimulation on Physiologic and Developmental Responses of Premature Infants in Neonatal Intensive Care." *Pediatric Nursing* 24, no. 6 (1998): 532–38.

Standley, Jayne M. "The Effects of Contingent Music to Increase Non-Nutritive Sucking of Premature Infants." *Pediatric Nursing* 26, no. 5 (1998): 493.

Steen, Francis. "A Cognitive Account of Aesthetics." In *The Artful Mind. Cognitive Science and the Riddle of Creativity*, edited by Mark Turner, 57–72. Oxford: Oxford University Press, 2006.

Steinig, Wolfgang. *Als die Wörter tanzen lernten. Ursprung und Gegenwart von Sprache.* München: Elsevier, 2007.

Sterelny, Kim. "Language, Gesture, Skill: The Co-Evolutionary Foundations of Language." *Philosophical Transactions of the Royal Society B: Biological Sciences* 367, no. 1599 (2012): 2141–51. Accessed August 10, 2017. doi:10.1098/rstb.2012.0116.

Storey, Robert. *Mimesis and the Human Animal: On the Biogenetic Foundations of Literary Representation*. Evanston, IL: Northwestern University Press, 1996.

Strasser, Ingrid. *Bedeutungswandel und strukturelle Semantik. 'Marotte, Laune, Tick im literarischen Deutsch der Gegenwart und der Frühen Goethezeit.* Wien: Karl M. Halosar, 1976.

Stumpf, Carl. "Musikpsychologie in England. Betrachtungen über die Herleitung der Musik aus der Sprache und aus dem thierischen Entwickelungsproceß, über Empirismus und Nativismus in der Musiktheorie." *Vierteljahrsschrift für Musikwissenschaft* 1 (1885): 261–349.

Sturgess, Philip J. M. *Narrativity. Theory and Practice*. Oxford: Clarendon, 1992.

Sugiyama, Michelle S. "On the Origins of Narrative: Storyteller Bias as a Fitness-Enhancing Strategy." *Human Nature* 7, no. 4 (1996): 403. Accessed July 28, 2017. doi:10.1007/bf02732901.

Sugiyama, Michelle S. "Reverse-Engineering Narrative: Evidence of Special Design." In *The Literary Animal: Evolution and the Nature of Narrative*, edited by Jonathan Gottschall and David Sloan Wilson, 177–97. Evanston, IL: Northwestern University Press, 2009.

Symons, Donald. "Beauty Is in the Adaptations of the Beholder: The Evolutionary Psychology of Human Female Sexual Attractiveness." In *Sexual Nature, Sexual Culture,* edited by Paul R. Abramson and Stefen D. Pinkerton, 80–120. Chicago: University of Chicago Press, 1995.

Tamir, Diana I., Andrew B. Bricker, David Dodell-Feder, and Jason P. Mitchell. "Reading Fiction and Reading Minds: The Role of Simulation in the Default Network." *Social Cognitive and Affective Neuroscience* 11, no. 2 (2016): 215–24. Accessed July 14, 2017. doi:10.1093/scan/nsv114.

Tattersall, Ian. "Human Origins: Out of Africa." *Proceedings of the National Academy of Sciences of the USA* 106 (2009): 16018–20.

Thornhill, Randy. "Darwinian Aesthetics." In *Handbook of Evolutionary Psychology: Ideas, Issues and Applications*, edited by Charles Crawford, Dennis Krebs, and Mahwah L., 543–72. London: Erlbaum, 1998.

Thornhill, Randy, and Steven W. Gangestad. "Human Facial Beauty. Averageness, Symmetry, and Parasite Resistance." *Human Nature* 4, no. 3 (1993): 237–69. Accessed July 14, 2017. doi:10.1007/BF02692201.

Tiggemann, Marika, and Sarah J. Kenyon. "The Hairlessness Norm: The Removal of Body Hair in Women." *Sex Roles* 39, no. 11 (1998): 873–85. Accessed July 17, 2017. doi:10.1023/a:1018828722102.

Tiggemann, Marika, and Christine Lewis. "Attitudes toward Women's Body Hair: Relationship with Disgust Sensitivity." *Psychology of Women Quarterly* 28, no. 4 (2004): 381–87. Accessed August 22, 2017. doi:10.1111/j.1471-6402.2004.00155.x.

Tigges, Margarete. "Farbbevorzugungen bei Fischen und Vögeln." *Zeitschrift für Tierpsychologie* 20 (1963): 129–42.

Tinbergen, Nikolaas. "On Aims and Methods of Ethology." *Zeitschrift für Tierpsychologie* 20 (1963): 410–33.

Todd, Peter. "Simulating the Evolution of Musical Behavior." In *The Origins of Music*, edited by Nils L. Wallin, Björn Merker, and Steven Brown, 361–88. Cambridge, MA: The MIT Press, 2000.

Toerien, Merran, and Sue Wilkinson. "Gender and Body Hair: Constructing the Feminine Woman." *Women's Studies International Forum* 26, no. 4 (2003): 333–44.

Tomasello, Michael. *Cultural Origins of Human Cognition*. Cambridge, MA: Harvard University Press, 1999.

Tomasello, Michael. *Origins of Human Communication*, Cambridge, MA: The MIT Press, 2008.

Tooby, John, and Leda Cosmides. "Does Beauty Build Adapted Minds? Toward an Evolutionary Theory of Aesthetics, Fiction and the Arts." *SubStance. A Review of Theory and Literary Citicism* 30 (= Special Issue: *On the Origin of Fictions: Interdisciplinary Perspectives*) (2001): 6–27.

Trehub, Sandra E., Anna M. Unyk, and Laurel J. Trainor. "Adults Identify Infant-Directed Music across Cultures." *Infant Behavior and Development* 16, no. 2 (1993): 193–211.

Trehub, Sandra E., David S. Hill, and Stuart B. Kamenetsky. "Parents' Sung Performances for Infants." *Canadian Journal of Experimental Psychology* 51, no. 4 (1997): 385–96.

Trehub, Sandra E., E. Glenn Schellenberg, and Stuart B. Kamenetsky. "Infants' and Adults' Perception of Scale Structure." *Journal of Experimental Psychology: Human Perception and Performance* 25, no. 4 (1999): 965–75. Accessed August 22, 2017. doi:10.1037/0096-1523.25.4.965.

Trehub, Sandra. "Human Processing Predispositions and Musical Universals." In *The Origins of Music*, edited by Nils L. Wallin, Björn Merker and Steven Brown, 427–48. Cambridge, MA: The MIT Press, 2000.

Trinkaus, Erik. *The Shanidar Neandertals*. New York: Academic Press, 1983.

Trivers, Robert L. "Parental Investment and Sexual Selection." In *Sexual Selection and the Descent of Man 1871-1971*, edited by Bernard Campbell, 136–79. London: Heineman, 1971.

Trivers, Robert L. "Deceit and Self-Deception." In *Mind the Gap. Tracing the Origin of Human Universals*, edited by Peter M. Kappeler and Joan B. Silk, 373–94. Berlin: Springer, 2010.

Tschacher, Wolfgang, Steven Greenwood, Volker Kirchberg, Stéphanie Wintzerith, Karen van den Berg, and Martin Tröndle. "Physiological Correlates of Aesthetic Perception of Artworks in a Museum." *Psychology of Aesthetics, Creativity, and the Arts* 6, no. 1 (2012): 96–103. Accessed July 12, 2017. doi:10.1037/a0023845.

Turner, Frederick, and Ernst Pöppel. "The Neural Lyre: Poetic Meter, the Brain, and Time." *Poetry* 142, no. 5 (1983): 277–309. Accessed August 10, 2017. doi:10.2307/20599567.

Turner, Frederick. "The Sociobiology of Beauty." In *Sociobiology and the Arts*, edited by Baptist Bedaux and Brett Cooke, 63–82. Amsterdam: Editions Rodopi, 1999.

Turner, Mark. *The Literary Mind*. Oxford: Oxford University Press, 1996.

Turner, Mark. "The Cognitive Study of Art, Language, and Literature." Poetics Today 23 (2002): 9–20.

Turner, Mark. "The Art of Compression." In *The Artful Mind. Cognitive Science and the Riddle of Human Creativity*, edited by Mark Turner, 93–114. New York: Oxford University Press, 2006.

Turner, Victor. *From Ritual to Theatre. The Human Seriousness of Play*. New York: Performing Arts Journal Publications, 1982.

Unyk, Anna M., Sandra E. Trehub, Laurel J. Trainor, and E. Glenn Schellenberg. "Lullabies and Simplicity: A Cross-Cultural Perspective." *Psychology of Music* 20, no. 1 (1992): 15–28.

Van de Cruys, Sander, and Johan Wagemans. "Putting Reward in Art: A Tentative Prediction Error Account of Visual Art." *i-Perception* 2, no. 9 (2011): 1035-1062. Accessed July 28, 2017. doi: 10.1068/i0466aap.

Vanhaeren, Marian, Francesco Errico, Chris Stringer, Sarah L. James, Jonathan A. Todd, and Henk K. Mienis. "Middle Paleolithic Shell Beads in Israel and Algeria." *Science* 312, no. 5781 (2006): 1785–88.

Van Schaik, Carel P., Arnold F. Sitompul, and Elizabeth A. Fox. "Manufacture and Use of Tools in Wild Sumatran Orangutans. Implications for Human Evolution." *Naturwissenschaften* 83, no. 4 (1996): 186–88.

Vaughn, Kathryn. "Music and Mathematics: Modest Support for the Soft-Claimed Relationship." *The Journal of Aesthetic Education* 34 (2000): 149–66. Accessed July 27, 2017. doi:10.2307/3333641.

Veblen, Thorstein. *The Theory of the Leisure Class. An Economic Study of Institutions*. New York: The MacMillan Company, 1912.

Vega, Manuel de, Immaculada Léon, and José M. Diaz. "The Representation of Changing Emotions in Reading Comprehension." *Cognition and Emotion* 10, no. 3 (1996): 303–22. Accessed September 1, 2017. doi:10.1080/026999396380268.

Verpooten, Jan. *Art and Signaling in a Cultural Species*. PhD diss, KU Leuven/ University of Antwerp, 2015.

Vitouch, Oliver, and Olivia Ladinig. "Preface." *Musicae Scientiae* 13, no. 2 supplement (2009): 7-11.

Voland, Eckart, and Karl Grammer, eds. *Evolutionary Aesthetics*. Heidelberg: Springer, 2003.

Voland, Eckart. "Aesthetic Preferences in the World of Artifacts – Adaptations for the Evaluation of 'Honest Signals'?" In *Evolutionary Aesthetics*, edited by Voland Eckart and Karl Grammer, 239–60. Heidelberg: Springer, 2003.

Vorderer, Peter, Francis F. Sterne, and Elaine Chan. "Motivation." In *Psychology of Entertainment*, edited by Jennings Bryant and Peter Vorderer, 3–18. New York: Routledge, Taylor & Francis Group, 2011.

Wagner, Valentin, Winfried Menninghaus, Julian Hanich, and Thomas Jacobsen. "Art Schema Effects on Affective Experience: The Case of Disgusting Images." *Psychology of Aesthetics, Creativity, and the Arts* 8, no. 2 (2014): 120–29. Accessed December 17, 2014. doi:10.1037/a0036126.

Wallot, Sebastian, & Menninghaus, Winfried. "Ambiguity Effects of Rhyme and Meter." *Journal of Experimental Psychology: Learning, Memory, and Cognition* 44, no. 12 (2018), 1947–1954. doi:10.1037/xlm0000557.

Walton, Kendall L. *Mimesis as Make-Believe. On the Foundations of the Representational Arts*. Cambridge, MA: Harvard University Press, 1993.

Washburn, Sherwood L. "Comment On: 'A Possible Evolutionary Basis for Aesthetic Appreciation in Men and Apes.'" *Evolution* 24, no. 4 (1970): 824–25. Accessed September 1, 2017. doi:10.1111/j.1558-5646.1970.tb01818.x.

Watts, Ian. "Ochre in the Middle Stone Age of Southern Africa: Ritualised Display or Hide Preservative?" *South African Archaelogical Bulletin* 57, no. 175 (2002): 1–14.

Wegener, Claudia. *Medien, Aneignung und Identität. 'Stars' im Alltag jugendlicher Fans*. Wiesbaden: VS Verlag für Sozialwissenschaften, 2008.

Wegner, K. Mathias, Martin Kalbe, Joachim Kurtz, Thorsten B. H. Reusch, and Manfred Milinski. "Parasite Selection for Immunogenetic Optimality." *Science* 301, no. 5638 (2003): 1343.

Weidlé, Wladimir. *Gestalt und Sprache des Kunstwerks. Studien zur Grundlegung einer nichtästhetischen Kunsttheorie*. Mittenwald: Mäander Kunstverlag, 1981.

Welsch, Wolfgang. "Animal Aesthetics." *Contemporary Aesthetics* 2 (2004).

Werner, Gregory M., and Peter M. Todd. "Too Many Love Songs: Sexual Selection and the Evolution of Communication." In *Fourth European Conference on Artificial Life*, edited by Phil Husbands and Inman Harvey, 434–43. Cambridge, MA: The MIT Press, Bradford Books, 1997.

Whissell, Cynthia. "Mate Selection in Popular Women's Fiction." *Human Nature* 7, no. 4 (1996): 427–47. Accessed September 1, 2017. doi:10.1007/bf02732902.

Wickler, Wolfgang. "Socio-Sexual Signals and their Intra-Specific Imitation among Primates." In Primate *Ethology*, edited by Desmond Morris, 69–147. London: Weidenfeld & Nicholson, 1967.

Wilson, David S. *Darwin's Cathedral: Evolution, Religion and the Nature of Society*. Chicago, IL: University of Chicago Press, 2002.

Wilson, Edward O. *Sociobiology. The New Synthesis*. Cambridge, MA: Belknap Press of Harvard University Press, 1975.

Wilson, Edward O. "On Art." In *Biopoetics. Evolutionary Explorations in the Arts*, edited by Brett Cooke and Frederick Turner, 71–96. Lexington, KY: ICUS, 1999.

Wiltermuth, Scott S. and Chip Heath. "Synchrony and Cooperation." *Psychological Science* 20, no. 1 (2009): 1–5. Accessed September 1, 2017. doi:10.1111/j.1467-9280.2008.02253.x.

Winkielman, Piotr, Jamin Halberstadt, Tedra Fazendeiro, and Steve Catty. "Prototypes Are Attractive Because They Are Easy on the Mind." *Psychological Science* 17, no. 9 (2006): 799–806. Accessed September 1, 2017. doi:10.1111/j.1467-9280.2006.01785.x.

Winner, Ellen, and Monica Cooper. "Mute Those Claims: No Evidence (Yet) for a Causal Link between Arts Study and Academic Achievement." *Journal of Aesthetic Education* 34, no. 3/4 (2000): 11–75. Accessed September 1, 2017. doi:10.2307/3333637.

Wöbse, Hans Hermann. "Landschaftsästhetik – Gedanken zu einem zu einseitig verwendeten Begriff." *Landschaft + Stadt* 13, (1981): 152–60.

Wrangham, Richard W., and Dale Peterson. *Demonic Males: Apes and the Origin of Human Violence*. Boston: Houghton Mifflin, 1996.

Wray, Alison. "Protolanguage as a Holistic System for Social Interaction." *Language and Communication* 18 (1998): 47–67.

Wynn, Thomas. "Handaxe Enigmas." *World Archaeology* 27 (1995): 10–24. Accessed September 1, 2017. doi:10.1080/00438243.1995.9980290.

Young, Dudley. *Origins of the Sacred. The Ecstasies of Love and War*. Kent: Little Brown and Company, 1992.

Young, Kay, and Jeffrey L. Saver. "The Neurology of Narrative." *A Review of Theory and Literary Citicism* 30 (= Special Issue: *On the Origin of Fictions: Interdisciplinary Perspectives*) (2001): 72–84.

Zahavi, Amotz. "Mate Selection: A Selection for a Handicap." *Journal of Theoretical Biology* 53, no. 1 (1975): 205–14. Accessed September 4, 2017. doi:10.1016/0022-5193(75)90111-3.

Zahavi, Amotz. "Decorative Patterns and the Evolution of Art." *New Scientist* 80 (1978): 182–84.

Zahavi, Amotz, and Avishag Zahavi. *The Handicap Principle: A Missing Piece of Darwin's Puzzle*. New York: Oxford University Press, 1997.

Zajonc, Robert B. "Attitudinal Effects of Mere Exposure." *Journal of Personality and Social Psychology* 9, no. 2 (1968): 1–27. Accessed August 10, 2017. doi:10.1037/h0025848.

Zeki, Semir. *Inner Vision: An Exploration of Art and the Brain*. Oxford: Oxford University Press, 1999.

Zeki, Semir. "The Neurology of Ambiguity." In *The Artful Mind. Cognitive Science and the Riddle of Creativity*, edited by Mark Turner, 243–70. Oxford: Oxford University Press, 2006.

Zentner, Marcel, Didier Grandjean, and Klaus R. Scherer. "Emotions Evoked by the Sound of Music: Characterization, Classification and Measurement." *Emotion* 8, no. 4 (2008): 494–521. Accessed August 10, 2017. doi:10.1037/1528-3542.8.4.494.

Ziehen, Theodor. *Vorlesungen über Ästhetik*. 2 vols. Halle an der Saale: Niemeyer, 1923 and 1925.

Zilhão, João, Diego E. Angelucci, Ernestina Badal-García, Francesco d'Errico, Floréal Daniel, Laure Dayet, Katerina Douka, Thomas F. G. Higham, María José Martínez-Sánchez, Ricardo Montes-Bernárdez, Sonia Murcia-Mascarós, Carmen Pérez-Sirvent, Clodoaldo Roldán-García, Marian Vanhaeren, Valentín Villaverde, Rachel Wood, and Josefina Zapata. "Symbolic Use of Marine Shells and Mineral Pigments by Iberian Neandertals." *Proceedings of the National Academy of Sciences* 107, no. 3 (2010): 1023–28.

Zuckert, Rachel. *Kant on Beauty and Biology: An Interpretation of the, 'Critique of Judgment.'* London: Cambridge University Press, 2007.

Zunshine, Lisa. *Why We Read Fiction? Theory of Mind and the Novel*. Columbus: Ohio State University Press, 2006.

Zwaan, Rolf A. "Some Parameters of Literary and News Comprehension: Effects of Discourse-Type Perspective on Reading Rate and Surface Structure Representation." *Poetics* 20, no. 2 (1991): 139–56. Accessed August 10, 2017. doi:10.1016/0304-422X(91)90003-8.

Index

www.ingramcontent.com/pod-product-compliance
Ingram Content Group UK Ltd.
Pitfield, Milton Keynes, MK11 3LW, UK
UKHW022315300426
470551UK00008B/123

9 781644 696101